SpringerBriefs in Optimization

Series Editors

Sergiy Butenko, Texas A&M University, College Station, TX, USA
Mirjam Dür, University of Trier, Trier, Germany
Panos M. Pardalos, University of Florida, Gainesville, FL, USA
János D. Pintér, Lehigh University, Bethlehem, PA, USA
Stephen M. Robinson, University of Wisconsin-Madison, Madison, WI, USA
Tamás Terlaky, Lehigh University, Bethlehem, PA, USA
My T. Thai ⓘ, University of Florida, Gainesville, FL, USA

SpringerBriefs in Optimization showcases algorithmic and theoretical techniques, case studies, and applications within the broad-based field of optimization. Manuscripts related to the ever-growing applications of optimization in applied mathematics, engineering, medicine, economics, and other applied sciences are encouraged.

More information about this series at http://www.springer.com/series/8918

Dimitris Souravlias • Konstantinos E. Parsopoulos
Ilias S. Kotsireas • Panos M. Pardalos

Algorithm Portfolios

Advances, Applications, and Challenges

Dimitris Souravlias
Logistics Management Department
Helmut-Schmidt University
Hamburg
Hamburg, Germany

Konstantinos E. Parsopoulos
Department of Computer Science &
Engineering
University of Ioannina
Ioannina, Greece

Ilias S. Kotsireas
Department of Physics & Computer Science
Wilfrid Laurier University
Waterloo, ON, Canada

Panos M. Pardalos ⓘ
Industrial and Systems Engineering
University of Florida
Gainesville, FL, USA

ISSN 2190-8354   ISSN 2191-575X   (electronic)
SpringerBriefs in Optimization
ISBN 978-3-030-68513-3   ISBN 978-3-030-68514-0   (eBook)
https://doi.org/10.1007/978-3-030-68514-0

Mathematics Subject Classification: 90C26, 68Txx, 62-07

This Springer imprint is published by the registered company Springer Nature Switzerland AG
The registered company address is: Gewerbestrasse 11, 6330 Cham, Switzerland

To my parents, Nikolaos and Aikaterini, and my partner, Anastasia

D. Souravlias

To my shining stars, Vangelis and Manos

K.E. Parsopoulos

To my parents, Sotirios and Olympia

I.S. Kotsireas

To my parents, Miltiades and Kalypso

P.M. Pardalos

Preface

In the era of artificial intelligence and data-driven decisions, optimization is placed in a salient position among relevant mathematical and computer science tools. In every second that passes, billions of electronic devices around the world run algorithms that optimize something. For instance, it can be the electricity consumption in our office, a portfolio of trading stocks, the driving distance between two points on a city map, or the inventory of our business warehouse.

A first and necessary step in solving optimization problems is the development of a model that describes the physical problem as accurately as possible. Experience of many decades has shown that it is very convenient to model the physical problem as the minimization (or maximization) of a suitable objective function. This way, the critical points of the objective function can be translated to candidate solutions of the physical problem.

In some cases, the derived models are simple and can be tackled even analytically. Unfortunately, this is rarely the case in real-world applications where the analytical form of the objective function may become quite complex. Nonlinear and combinatorial optimization problems have traditionally constituted intriguing challenges for the practitioners. As their size increases, they require efficient procedures for probing huge domains or function landscapes with numerous minimizers. Additional issues that can exponentially increase their difficulty include non-analytical functional forms, discontinuities, non-differentiability, and noisy objective values.

Simulation-based optimization offers another category of such demanding problems. They usually appear in contemporary applications where knowledge is extracted from data, and objective values are derived through simulation procedures rather than evaluating mathematical formulae. Efficient algorithms combined with high-end hardware are typically required for successfully engaging such problems.

From early mathematical approaches of the previous century up to modern metaheuristics and high-performance optimization algorithms, researchers have continuously striven to exploit special characteristics of optimization problems in order to develop efficient solvers. The abundance of optimization problem types has lead to the rich algorithmic artillery of our days.

A question that has steadily surfaced in researchers' minds refers to the existence of a universal algorithm that could possibly outperform all other optimization algorithms over all possible problems and performance metrics. The series *No Free Lunch Theorems* [2, 166] has proved that such algorithm can hardly exist. However, numerous experimental findings suggest that by narrowing our considered problem type down to very specific forms (e.g., strongly convex and multimodal), we can clearly distinguish some algorithms against others. Thus, traits of free lunch appear when specific problem types are considered. This is an ongoing motivation for keeping research in optimization algorithms active.

The selection of a specific category of algorithms for a given optimization problem is typically driven by the practitioner's previous experience. Successful algorithms for almost all problem types are reported in the relevant literature, supported by numerous experimental results. Nevertheless, selecting the best one among them is frequently a puzzling decision. This is due to the variety of characteristics and peculiarities of each algorithm.

Quite often, optimization algorithms possess different properties in terms of efficiency, effectiveness, parallelization capability, implementation simplicity, mathematical requirements of the objective function, parametrization effort, hardware-failure tolerance, and exploitation-vs-exploration trade-off, among others. Even the phase of the optimization procedure plays a role in performance, with some algorithms performing well in early iterations while others performing better at the later iterations of a run.

Sometimes the number of possible algorithm-selection options is overwhelming, and the final decision is solely based on the practitioner's background and experience. In order to overcome the risk of choosing an algorithm that systematically fails or has inferior performance on a given problem, synergism or competition among multiple algorithms has been promoted as a promising solution. In both cases, a number of algorithms is selected instead of a single one, and they are harnessed under schemes based on either their cooperation (in terms of solution findings) or their competition for the available computation resources.

While cooperation assumes that each algorithm has reserved a minimum number of resources, competition promotes the winner-takes-all concept, that is, algorithms are subsequently eliminated until only one is left running. Obviously, the former approach aims at deriving possible benefits from all algorithms through resource sharing, especially if communication among them is allowed, while the latter aims at distinguishing only the most promising approach, which eventually claims all the resources.

Algorithm portfolios stand on the cooperation-oriented side. They were introduced in the pioneering work of B.A. Huberman, R.M. Lukose, and T. Hogg [68] as a type of algorithm ensemble where each algorithm is assigned a portion of the available computational resources. The algorithms were assumed to run interchangeably on a single processor, spending their specific portion of computation time in batches. The analysis in [68] showed there are resource-allocation plans that are associated with higher probability of solving the problem at hand if more than one algorithm is executed, instead of assigning the whole computation budget on a single algorithm.

Soon after the inaugural work of Huberman et al., research on algorithm portfolios has grown. The use of metaheuristics as well as communication among the algorithms have been considered. Relevance with ongoing problems such as the algorithm selection problem in artificial intelligence and the automatic resource allocation and planning have been investigated. New applications appeared in diverse fields.

The authors of the present book have contributed a number of works in this direction. They have worked on metaheuristics-based algorithm portfolios with sophisticated resource allocation schemes. Also, they have implemented parallel algorithm portfolios with interaction among the constituent algorithms. The derived approaches have been demonstrated to be effective on various applications and constitute the backbone of the present book.

Suggested Audience and Outline of the Book

The present monograph offers a concise introduction in algorithm portfolios. Basic ideas, implementation issues, specialized resource allocation schemes, and corresponding applications based on the authors' research experience are presented, along with recent advances and future challenges.

The book is organized in the following chapters:

- **Chapter 1** is devoted to metaheuristics. This is an introductory chapter that offers brief descriptions of algorithms from both categories of trajectory-based and population-based metaheuristics. The chapter exposes the non-expert reader to the basics of specific metaheuristics that will be later used for the construction of algorithm portfolios. To this end, tabu search, variable neighborhood search, and iterated local search are presented from the class of trajectory-based methods, while particle swarm optimization, differential evolution, and an enhanced differential evolution variant come from the class of population-based methods.
- **Chapter 2** introduces algorithm portfolios in a formal context. The presentation underlines the key concepts and design challenges. Integral components and inherent mechanisms are presented, along with related concepts. These include the selection of constituent algorithms, allocation of the available computation resources, and topics regarding sequential and parallel application, which are further analyzed in subsequent chapters.
- **Chapter 3** outlines methods for selecting the constituent algorithms of the portfolio, a problem related to the algorithm selection challenge in artificial intelligence. Basic approaches are reported, including feature-based and statistical selection procedures that have already been used for the construction of algorithm portfolios.
- **Chapter 4** immerses the reader in the core of the algorithm portfolio engine, that is, the resource allocation mechanism. Recent sophisticated resource allocation schemes are thoroughly analyzed. These include the recent trading-based allocation mechanism, allocation based on performance forecasting, and adaptive online allocation.

- **Chapter 5** discusses parallelization of algorithm portfolios, a critical issue for efficient implementations in difficult problems. The main differences between sequential and parallel models are outlined and various models of both types are reviewed.
- **Chapter 6** reports three recent applications of algorithm portfolios in the fields of combinatorics, production planning, and humanitarian logistics. The applications are discussed in detail, in order to expose the design decisions required to successfully apply an algorithm portfolio.
- **Epilogue** closes the book, providing a summary of the presented material and directions for future research.

The book is expected to be useful for researchers and doctoral students in relevant domains that seek a quick exposure of the field or a primary reference point. Having focused mostly on the applicability of the methods, we insisted on algorithmic presentation and clarity rather than theoretical details. Thus, the non-expert reader is expected to find this book useful for starting designing and implementing algorithm portfolios.

Hamburg, Germany   Dimitris Souravlias

Ioannina, Greece   Konstantinos E. Parsopoulos

Waterloo, ON, Canada   Ilias S. Kotsireas

Gainesville, FL, USA   Panos M. Pardalos

July 2020

Acknowledgments

K.E. Parsopoulos is grateful to Professors Michael N. Vrahatis, Isaac E. Lagaris, and Ilias S. Kotsireas for sharing their deep knowledge, as well as for their unlimited support and friendship. Deep respect goes to Professor Panos M. Pardalos for his huge contribution in the field of computational optimization, which has been a great source of inspiration and motivation. Special thanks go to Professor Konstantina Skouri for our fruitful collaboration in the field of operations research.

P.M. Pardalos was supported by the Paul and Heidi Brown Preeminent Professorship at ISE, University of Florida (USA), and a Humboldt Research Award (Germany).

Contents

Chapter 1
Metaheuristic Optimization Algorithms

In this introductory chapter, specific state-of-the-art metaheuristic optimization algorithms are outlined. The presented algorithms belong to the broad trajectory-based and population-based categories, and they are particularly selected as they constitute the building blocks of algorithm portfolios presented in the forthcoming chapters.

1.1 Introduction

Metaheuristics can be distinguished in two main categories, namely, trajectory-based (or local search) and population-based [109]. Trajectory-based refers to approaches that use a single search point that is iteratively updated through the application of neighborhood-search operators. Frequently, such algorithms incorporate hill-climbing mechanisms that render them capable of visiting a multitude of minimizers. Population-based refers to algorithms that employ a population of search points, also referred to as candidate solutions. The collective dynamic of the population is based on interactions among its members. These interactions promote search in the most promising regions of the considered domain. Inherent global search and parallelization capabilities are typical characteristics of this algorithm type [128].

For presentation purposes, we henceforth assume that the problem under consideration is the bound-constrained minimization problem defined as

$$\min_{\mathbf{x} \in \mathcal{X}} f(\mathbf{x}), \tag{1.1}$$

where $\mathcal{X}$ is the n-dimensional bounded search space under consideration. The forthcoming trajectory-based approaches have been originally introduced as discrete optimization solvers. Thus, despite their extensions for continuous problems, it is

© The Author(s), under exclusive license to Springer Nature Switzerland AG 2021
D. Souravlias et al., *Algorithm Portfolios*, SpringerBriefs in Optimization,
https://doi.org/10.1007/978-3-030-68514-0_1

more straightforward to consider X as a discrete search space in those cases. On the other hand, the population-based approaches that will be later presented have been originally designed for continuous problems. Therefore, it is natural to consider X as a subset of $\mathbb{R}^n$ in those cases.

Algorithm 1 – Tabu search

Input: Search space X; tabu list size l
Output: Best detected solution $\mathbf{x}^*$

1: $t \leftarrow 0$
2: $L \leftarrow \emptyset$
3: $\mathbf{x}^{(t)} \leftarrow$ **initialization**(X)
4: $\mathbf{x}^* \leftarrow \mathbf{x}^{(t)}$
5: **repeat**
6:     $\mathbf{x}^{(t+1)} \leftarrow \underset{\mathbf{y} \in N_{\mathbf{x}^{(t)}} \setminus L}{\arg \min} \ f(\mathbf{y})$
7:     $L \leftarrow L \cup \{\mathbf{x}^{(t+1)}\}$
8:     **if** $(|L| > l)$ **then**
9:         **remove** $-$ **oldest** $-$ **entry**(L)
10:     **end if**
11:     $\mathbf{x}^* \leftarrow$ **update** $-$ **best** $(\mathbf{x}^*, \mathbf{x}^{(t+1)})$
12:     $t \leftarrow t + 1$
13: **until** (termination criterion is satisfied)
14: **return** $\mathbf{x}^*$

Modifications to adapt each algorithm from one problem type to another can be easily found in the relevant literature. In the pseudocodes presented in the following paragraphs, the `rand()` function stands for the random number generator providing decimal numbers in the interval $[0, 1]$.

1.2   Trajectory-Based Metaheuristics

Trajectory-based metaheuristics are among the best-studied optimization algorithms. They have been individually applied on both discrete and continuous problems, and their quality has been verified in multidisciplinary optimization tasks [52]. Tabu search, variable neighborhood search, and iterated local search are considered to be among the most popular approaches of this type. As we will later see, they have been successfully used to compose algorithm portfolios especially for discrete optimization problems. In the following sections, we offer brief algorithmic descriptions of these approaches, and particularly, the variants that have been used in the algorithm portfolios are presented in later chapters.

1.2.1 *Tabu Search*

Tabu search is an established metaheuristic that has been applied on numerous discrete optimization problems, spanning different application fields [51, 56]. First, the algorithm samples new points in the local neighborhood of the current candidate solution. Then, it moves to the best sampled point (possibly the best point of the whole neighborhood) regardless of its relevant quality compared to the current candidate solution. This feature equips tabu search with the hill-climbing property, which allows it to visit the neighborhoods of multiple minimizers.

Algorithm 2 – Variable neighborhood search

Input: Search space X; neighborhood types N_k, $k = 1, \ldots, k_{max}$
Output: Best detected solution $\mathbf{x}^*$

1: $\mathbf{x} \leftarrow$ **initialization**(X)
2: $\mathbf{x}^* \leftarrow \mathbf{x}$
3: **repeat**
4:     $k \leftarrow 1$
5:     **while** ($k \leqslant k_{max}$) **do**
6:         $\mathbf{x}' \leftarrow$ **perturb** $(\mathbf{x}, N_k)$
7:         **repeat**
8:             $\mathbf{x}'' \leftarrow \mathbf{x}'$
9:             $\mathbf{x}' \leftarrow \arg\min_{z \in N_k(\mathbf{x}')} f(z)$
10:        **until** $\left(f(\mathbf{x}'') \leqslant f(\mathbf{x}') \right)$
11:        **if** $\left(f(\mathbf{x}'') < f(\mathbf{x}) \right)$ **then**
12:            $\mathbf{x} \leftarrow \mathbf{x}''$
13:            $k \leftarrow 1$
14:            $\mathbf{x}^* \leftarrow$ **update** $-$ **best** $(\mathbf{x}^*, \mathbf{x})$
15:        **else**
16:            $k \leftarrow k + 1$
17:        **end if**
18:     **end while**
19: **until** (termination criterion is satisfied)
20: **return** $\mathbf{x}^*$

In order to avoid exploring already visited regions of the search space, the algorithm is equipped also with a short-term memory structure of predefined size, called the tabu list. A number of the most recently visited positions are stored in this list, and they are excluded from the possible new moves of the algorithm. Each new point accepted by the algorithm is automatically inserted in the list, and if the list is already full, the oldest entry is removed.

Despite the hill-climbing mechanism of tabu search, there is always the risk of confining the search in narrow regions of the search space. For this reason, the algorithm is usually applied within a multistart framework. Thus, if the best solution cannot be improved for a number of iterations, T_{noimp}, the algorithm is re-initialized to a new (random) candidate solution.

Similarly to other metaheuristics, a run of tabu search is terminated when a predefined computational budget is exceeded or the best solution cannot be further improved. Detailed descriptions of tabu search can be found in [51, 54–56], and a general-purpose pseudocode is provided in Algorithm 1.

Algorithm 3 – Iterated local search
_____
Input: Search space $\mathcal{X}$; local search procedure H
Output: Best detected solution $\mathbf{x}^*$
_____
1: $\mathbf{x} \leftarrow$ **initialization**$(\mathcal{X})$
2: $\mathbf{x}^* \leftarrow \underset{\mathbf{y} \in N(\mathbf{x}) \cap \mathcal{X}}{\arg\min} \ f(\mathbf{y})$
3: **repeat**
4:    $\mathbf{x}' \leftarrow$ **perturb** $(\mathbf{x}^*)$
5:    $\mathbf{x}'' \leftarrow \underset{\mathbf{y} \in N(\mathbf{x}') \cap \mathcal{X}}{\arg\min} \ f(\mathbf{y})$
6:    $\mathbf{x}^* \leftarrow$ **update** $-$ **best** $(\mathbf{x}^*, \mathbf{x}'')$
7: **until** (termination criterion is satisfied)
8: **return** $\mathbf{x}^*$
_____

1.2.2   Variable Neighborhood Search

Variable neighborhood search is a trajectory-based metaheuristic primarily designed to tackle combinatorial optimization problems [99]. It has been used in numerous application fields, scoring tremendous success in operations research problems such as vehicle routing [43], order batching [97], and plant location [63].

A significant number of variants have been hitherto proposed in the literature [65]. We consider a rather generic scheme that combines local search with a perturbation mechanism, both applied within a multi-neighborhood framework that interchangeably considers a number of predetermined neighborhood types.

The algorithm is initiated with a randomly generated position $\mathbf{x} \in \mathcal{X}$ and its first neighborhood type N_1. The current candidate solution is perturbed in N_1, producing a new point $\mathbf{x}'$ that initiates a local search procedure in the current neighborhood. The outcome of the local search is a new point $\mathbf{x}''$. If this point improves $\mathbf{x}$, then it is selected as the new current candidate solution, and the neighborhood index is reset to 1. Otherwise, the next neighborhood is selected, and a new perturbed point $\mathbf{x}'$ is sampled in it. The procedure is repeated until all neighborhoods have been used without success. In that case, given that computational budget is still available, the algorithm resets neighborhood index to 1 and starts over again.

Eventually, the algorithm is terminated when the available computation budget is exceeded or the best detected solution cannot be improved for a large number of iterations. Additionally, it is also restarted after a number of non-improving iterations. The pseudocode of the presented variable neighborhood search variant is reported in Algorithm 2. The reader is referred to [64] for a thorough presentation.

1.2.3   Iterated Local Search

Iterated local search has introduced a simple optimization framework for tackling discrete problems [92, 93]. The cornerstone of the algorithm is the iterative application of an embedded local search heuristic, combined with a special perturbation mechanism that offers essential exploration capabilities. Abundance of applications of the algorithm on a variety of scientific fields can be found in the relevant literature [33, 139, 177].

The algorithm starts from a random initial point $\mathbf{x}$, and applies a local search heuristic. The local search consists of subsequent steepest descent moves until a local minimizer is detected. The detected minimizer is perturbed in order to diversify the direction components of the incumbent position. Designing the perturbation mechanism is not a straightforward task because perturbation should not allow the algorithm to revert to the previous local minimizer or perform a random restart.

The perturbed point $\mathbf{x}'$ is fed to the employed local search procedure, which generates a new local minimizer $\mathbf{x}''$. Next, an acceptance procedure comes into play and probes if $\mathbf{x}''$ improves the best local minimizer so far. Finally, the best solution is passed to the next cycle of the algorithm, which repeats the same steps as above.

The generic framework of iterated local search is given in Algorithm 3. The reader is referred to [93] for a thorough presentation of the algorithm.

1.3   Population-Based Metaheuristics

In this section, two state-of-the-art population-based metaheuristics, namely, particle swarm optimization and differential evolution, coming from the broad fields of evolutionary computation and swarm intelligence are presented. Additionally, a differential evolution variant that differs from the typical algorithm is exposed. These algorithms originally target continuous optimization problems. They have gained increasing popularity in the past decades, exhibiting a vast amount of applications [41, 117]. Consequently, they can be considered as promising approaches for building algorithm portfolios for real-valued problems.

1.3.1   Particle Swarm Optimization

Particle swarm optimization is a well-studied population-based metaheuristic. Since its development in 1995 [74], numerous variants have been proposed and applied on a broad range of applications, spanning various scientific and technological fields [95, 117, 119, 125, 126, 156, 163].

The algorithm is based on the coordinated move of a population, called the swarm, of candidate solutions. Also, it takes advantage of both individual and collective findings, in order to identify the most promising regions of the search space. Let N be the number of candidate solutions (search points) that comprise the swarm:

$$S = \{\mathbf{x}_1, \mathbf{x}_2, \ldots, \mathbf{x}_N\}.$$

Algorithm 4 – Particle swarm optimization

Input: Swarm size N; parameters χ, c_1, c_2; computational budget
Output: Best detected solution $\mathbf{x}^*$

1: $t \leftarrow 0$
2: **for** $(i = 1 \ldots N)$ **do**
3: 　　$\left[\mathbf{x}_i^{(t)}, \mathbf{p}_i^{(t)}, \mathbf{v}_i^{(t)}\right] \leftarrow$ **initialization**$(\mathcal{X})$
4: 　　$\text{NB}_i \leftarrow$ **define** $-$ **neighborhood**$(\mathbf{x}_i)$
5: **end for**
6: $\mathbf{x}^* \leftarrow$ **update** $-$ **best** $\left(\mathbf{p}_1^{(t)}, \ldots, \mathbf{p}_N^{(t)}\right)$
7: **repeat**
8: 　　**for** $(i = 1 \ldots N)$ **do**
9: 　　　　$g_i \leftarrow$ **best** $-$ **index**(NB_i)
10: 　　　　**for** $(j = 1 \ldots n)$ **do**
11: 　　　　　　$\mathbf{v}_{ij}^{(t+1)} \leftarrow \chi\, \mathbf{v}_{ij}^{(t)} + c_1\, \text{rand}()\left(\mathbf{p}_{ij}^{(t)} - \mathbf{x}_{ij}^{(t)}\right) + c_2\, \text{rand}()\left(\mathbf{p}_{g_i j}^{(t)} - \mathbf{x}_{ij}^{(t)}\right)$
12: 　　　　　　$\mathbf{x}_{ij}^{(t+1)} \leftarrow \mathbf{x}_{ij}^{(t)} + \mathbf{v}_{ij}^{(t+1)}$
13: 　　　　**end for**
14: 　　**end for**
15: 　　**for** $(i = 1 \ldots N)$ **do**
16: 　　　　$\mathbf{p}_i^{(t+1)} \leftarrow$ **update** $-$ **best** $-$ **position** $\left(\mathbf{x}_i^{(t+1)}, \mathbf{p}_i^{(t)}\right)$
17: 　　**end for**
18: 　　$\mathbf{x}^* \leftarrow$ **update** $-$ **best** $\left(\mathbf{p}_1^{(t+1)}, \ldots, \mathbf{p}_N^{(t+1)}\right)$
19: 　　$t \leftarrow t + 1$
20: **until** (termination criterion is satisfied)
21: **return** $\mathbf{x}^*$

Each candidate solution $\mathbf{x}_i \in \mathcal{X} \subset \mathbb{R}^n$, also called a particle, is a n-dimensional vector

$$\mathbf{x}_i = (x_{i1}, x_{i2}, \ldots, x_{in})^\top \in \mathcal{X}$$

that iteratively updates its position in the search space, using an adaptable position shift, called velocity:

$$\mathbf{v}_i = (v_{i1}, v_{i2}, \ldots, v_{in})^\top.$$

Thus, at the $(t + 1)$-th iteration, the particles update their positions as follows:

$$\mathbf{x}_i^{(t+1)} = \mathbf{x}_i^{(t)} + \mathbf{v}_i^{(t+1)}, \quad i \in \{1, 2, \ldots, N\}. \tag{1.2}$$

Each initial position and velocity, $\mathbf{x}_i^{(0)}$ and $\mathbf{v}_i^{(0)}$, are taken randomly.

During its stochastic walk in the search space, each particle $\mathbf{x}_i$ stores in memory the best position, $\mathbf{p}_i$, it has ever visited, i.e., the one with the best objective value. Moreover, each particle $\mathbf{x}_i$ assumes a number of other particles, with which it exchanges information. This subset of the swarm is called the neighborhood of $\mathbf{x}_i$ and promotes the collective interaction and coordination of the particles. Although there is relevant tolerance on the formation of the neighborhoods, the most successful versions of the algorithm assume structured neighborhoods in the form of fully or partially connected graphs, also called neighborhood topologies [117]. The fully connected and the ring topology are among the most prominent ones [73].

The velocity update of each particle $\mathbf{x}_i$ takes into consideration its previous velocity, the best position discovered by the particle itself, and the best position discovered by its whole neighborhood. Thus, it is component-wisely defined as

$$v_{ij}^{(t+1)} = \chi \, v_{ij}^{(t)} + c_1 \, \texttt{rand}() \left(p_{ij}^{(t)} - x_{ij}^{(t)} \right) + c_2 \, \texttt{rand}() \left(p_{g_i j}^{(t)} - x_{ij}^{(t)} \right), \tag{1.3}$$

where $i = 1, 2, \ldots, N$ and $j = 1, 2, \ldots, n$. The parameters χ, c_1, and c_2, are used to control the swarm's dynamic. Their appropriate setting can ensure convergence of the swarm [37]. The index g_i indicates the best position achieved by any particle participating in the neighborhood of $\mathbf{x}_i$. Note that a different random number is used for each component update of $\mathbf{v}_i$ in order to avoid restricting the particles in subspaces of the search space $\mathcal{X}$.

Eventually, at each iteration, the best position of each particle competes against its new position, and it is updated as follows:

$$\mathbf{p}_i^{(t+1)} = \begin{cases} \mathbf{x}_i^{(t+1)}, & \text{if } f\left(\mathbf{x}_i^{(t+1)}\right) < f\left(\mathbf{p}_i^{(t)}\right), \\ \mathbf{p}_i^{(t)}, & \text{otherwise}, \end{cases} \tag{1.4}$$

Pseudocode of particle swarm optimization is provided in Algorithm 4. For a comprehensive survey of the algorithm and its applications, the reader is referred to [36, 117].

1.3.2 Differential Evolution

Differential evolution belongs to the broad category of evolutionary algorithms [145]. It is a population-based algorithm that applies deterministic difference-based operators for sampling the search space, while it is supported also by a stochastic recombination mechanism that permeates search stochasticity. A plethora of variants have been successfully used to solve (mostly) continuous optimization problems in a variety of application fields [41, 42].

Similarly to other evolutionary algorithms, differential evolution maintains a population of search points, also called individuals:

$$P = \{\mathbf{x}_1, \mathbf{x}_2, \ldots, \mathbf{x}_N\},$$

where $\mathbf{x}_i \in X \subset \mathbb{R}^n$. The population is randomly initialized. The algorithm proceeds by sampling a new point in the search space for each individual. For this purpose, it defines search directions by combining difference vectors between pairs of selected individuals from the population. Each sampled point is then stochastically recombined with its original individual to produce an offspring vector. The latter replaces the original vector in the population if it has better objective value.

Algorithm 5 – Differential evolution

Input: Swarm size N; parameters F, CR; computational budget
Output: Best detected solution $\mathbf{x}^*$

1: $t \leftarrow 0$
2: **for** $(i = 1 \ldots N)$ **do**
3:     $\mathbf{x}_i^{(t)} \leftarrow \textbf{initialization}(X)$
4: **end for**
5: $\mathbf{x}^* \leftarrow \textbf{update} - \textbf{best}\left(\mathbf{x}_1^{(t)}, \ldots, \mathbf{x}_N^{(t)}\right)$
6: **repeat**
7:     **for** $(i = 1 \ldots N)$ **do**
8:         $r_1, r_2 \leftarrow \textbf{random} - \textbf{index}(1 \ldots N), r_1 \neq r_2$
9:         $\mathbf{v}_i^{(t+1)} \leftarrow \mathbf{x}^* + F\left(\mathbf{x}_{r_1}^{(t)} - \mathbf{x}_{r_2}^{(t)}\right)$
10:        $j_{\text{rand}} \leftarrow \textbf{random} - \textbf{index}(1 \ldots n)$
11:        $\mathbf{u}_i^{(t+1)} \leftarrow \mathbf{x}_i^{(t+1)}$
12:        **for** $(j = 1 \ldots n)$ **do**
13:            **if** $(\texttt{rand}() \leqslant CR)$ OR $(j = j_{\text{rand}})$ **then**
14:                $u_{ij}^{(t+1)} \leftarrow v_{ij}^{(t+1)}$
15:            **end if**
16:        **end for**
17:    **end for**
18:    **for** $(i = 1 \ldots N)$ **do**
19:        $\mathbf{x}_i^{(t+1)} \leftarrow \textbf{select} - \textbf{best}\left(\mathbf{x}_i^{(t)}, \mathbf{u}_i^{(t+1)}\right)$
20:    **end for**
21:    $\mathbf{x}^* \leftarrow \textbf{update} - \textbf{best}\left(\mathbf{x}_1^{(t+1)}, \ldots, \mathbf{x}_N^{(t+1)}\right)$
22:    $t \leftarrow t + 1$
23: **until** (termination criterion is satisfied)
24: **return** $\mathbf{x}^*$

Putting it formally, each iteration t starts with the so-called mutation procedure. In this procedure, a new vector $\mathbf{v}_i^{(t+1)}$ is generated for each individual $\mathbf{x}_i^{(t)}$ by combining three or more individuals of the population. Different choices of the combined individuals define different mutation operators. In the relevant literature, five mutation operators have received considerable attention:

$$\text{DE1} : \mathbf{v}_i^{(t+1)} = \mathbf{x}_{\text{best}}^{(t)} + F\left(\mathbf{x}_{r_1}^{(t)} - \mathbf{x}_{r_2}^{(t)}\right), \tag{1.5}$$

$$\text{DE2} : \mathbf{v}_i^{(t+1)} = \mathbf{x}_{r_1}^{(t)} + F\left(\mathbf{x}_{r_2}^{(t)} - \mathbf{x}_{r_3}^{(t)}\right), \tag{1.6}$$

$$\text{DE3} : \mathbf{v}_i^{(t+1)} = \mathbf{x}_i^{(t)} + F\left(\mathbf{x}_{\text{best}}^{(t)} - \mathbf{x}_i^{(t)}\right) + F\left(\mathbf{x}_{r_1}^{(t)} - \mathbf{x}_{r_2}^{(t)}\right), \tag{1.7}$$

$$\text{DE4} : \mathbf{v}_i^{(t+1)} = \mathbf{x}_{\text{best}}^{(t)} + F\left(\mathbf{x}_{r_1}^{(t)} - \mathbf{x}_{r_2}^{(t)}\right) + F\left(\mathbf{x}_{r_3}^{(t)} - \mathbf{x}_{r_4}^{(t)}\right), \tag{1.8}$$

$$\text{DE5} : \mathbf{v}_i^{(t+1)} = \mathbf{x}_{r_1}^{(t)} + F\left(\mathbf{x}_{r_2}^{(t)} - \mathbf{x}_{r_3}^{(t)}\right) + F\left(\mathbf{x}_{r_4}^{(t)} - \mathbf{x}_{r_5}^{(t)}\right), \tag{1.9}$$

where $\mathbf{x}_{\text{best}}^{(t)}$ stands for the best individual of the population, i.e., the one with the best objective value, and $F \in [0, 2]$ denotes a user-defined parameter, also called the differential weight or mutation factor. This parameter is used to adjust the step size toward the search direction specified by the difference vectors. The indices

$$r_j \in \{1, 2, \ldots, N\} \setminus \{i\},$$

are randomly selected, and they are mutually different.

The mutation procedure is succeeded by crossover, which randomly blends the components of the original individual $\mathbf{x}_i^{(t)}$ and the mutant vector $\mathbf{v}_i^{(t+1)}$, producing a new offspring vector as follows:

$$u_{ij}^{(t)} = \begin{cases} v_{ij}^{(t+1)}, & \text{if } (\texttt{rand}() \leqslant CR), \text{ OR } (j = j_{\text{rand}}), \\ x_{ij}^{(t)}, & \text{otherwise}, \end{cases} \tag{1.10}$$

where $CR \in [0, 1]$ is a user-defined scalar parameter, also called the crossover probability, which controls the number of vector components inherited by the mutant vector to the offspring. Also, $j_{\text{rand}} \in \{1, 2, \ldots, n\}$ is a randomly selected index (individually selected for each vector $\mathbf{u}_i$), which ensures that at least one component of the offspring will be inherited from the mutant vector $\mathbf{v}_i^{(t+1)}$.

Finally, selection operator is applied to determine whether the produced vector $\mathbf{u}_i$ will replace the original individual $\mathbf{x}_i$:

$$\mathbf{x}_i^{(t+1)} = \begin{cases} \mathbf{u}_i^{(t+1)}, & \text{if } f\left(\mathbf{u}_i^{(t+1)}\right) < f\left(\mathbf{x}_i^{(t)}\right), \\ \mathbf{x}_i^{(t)}, & \text{otherwise}, \end{cases} \tag{1.11}$$

Special attention is needed in the appropriate configuration of the differential evolution algorithm, since numerical evidence suggests that its parameterization can have significant performance impact [171]. Algorithm 5 provides pseudocode of the algorithm with the DE1 operator (mild changes in lines 8 and 9 are needed for the rest of the operators), while a comprehensive survey along with various applications can be found in [42] and [41].

1.3.3 Enhanced Differential Evolution

Numerous modifications of the standard differential evolution algorithm have been introduced in the relevant literature [59, 98, 100, 101]. In this section, we present an enhanced differential evolution variant, originally proposed in [100]. This variant enhances the classical algorithm in three aspects:

(a) It defines a new mutation scheme.
(b) Crossover is based on the probabilistic selection between the new mutation scheme and the DE2 scheme of Eq. (1.6).
(c) It introduces a special restart strategy to successfully alleviate local minima.

The new mutation scheme is defined as

$$\mathbf{w}_i^{(t+1)} = \mathbf{x}_{r_1}^{(t)} + F_1 \left(\mathbf{x}_{\text{best}}^{(t)} - \mathbf{x}_{r_1}^{(t)} \right) + F_2 \left(\mathbf{x}_{r_1}^{(t)} - \mathbf{x}_{\text{worst}}^{(t)} \right), \tag{1.12}$$

where $\mathbf{x}_{r_1}^{(t)}$ is a randomly selected individual, $F_1, F_2 \in [0, 2]$ are called the differential weights, and $\mathbf{x}_{\text{best}}^{(t)}$ and $\mathbf{x}_{\text{worst}}^{(t)}$ denote the best and worst individual of the entire population at iteration t, respectively.

The enhanced crossover operator determines the trial vector as follows:

$$u_{ij}^{(t+1)} = \begin{cases} w_{ij}^{(t+1)}, & \text{if } (\text{cond} = 1) \text{ AND } \left(\text{rand}() \geqslant \frac{t_{\text{max}} - t}{t_{\text{max}}} \right), \\ v_{ij}^{(t+1)}, & \text{otherwise,} \end{cases} \tag{1.13}$$

where cond is the following condition:

$$\text{cond} \equiv (\text{rand}() \leqslant CR) \text{ OR } (j = j_{\text{rand}}),$$

and t_{max} is the total number of iterations. The rest of the parameters are identical to the standard differential evolution. Note that, in our case, v_{ij} is the j-th component of the mutation vector $\mathbf{v}_i$ produced through any of the Eqs. (1.5)–(1.9), and not only through Eq. (1.6), as suggested for the enhanced algorithm in [100].

Additionally, stagnation and/or premature convergence can be avoided by using a restart mechanism. We propose one that differs from the original in [100]. This mechanism is used for restarting each individual of the population except for the best one, which is kept unaltered. Specifically, restart applies mild perturbations to current individuals $\mathbf{x}_i$ resulting in perturbed individuals $\mathbf{x}'_i$, as follows:

$$x'_{ij} = x_{ij} \pm x_{\text{bias}}. \tag{1.14}$$

This scheme decreases the probability of getting trapped at local optima. Also, mild perturbations may enhance the local search capability of the algorithm. For each index j, the sign "+" or "−" in Eq. (1.14) is randomly chosen with equal probability. Also, the bias is defined as

$$x_{\text{bias}} = 1,$$

as this is the smallest step size in integer search spaces. Algorithm 5 with the appropriate modifications can be used to describe also the enhanced variant of differential evolution.

1.4 Synopsis

We outlined established metaheuristics of two major types, namely, trajectory-based and population-based. Trajectory-based algorithms iteratively probe the search space by applying local modifications to a single candidate solution, while hill-climbing capability allows them to escape from local minima. Prominent algorithms in this category are tabu search, variable neighborhood search, and iterated local search, which are mainly used for discrete optimization tasks.

On the other hand, population-based algorithms are based on the collective dynamic that arises from a population of interacting search agents. Two highly popular algorithms of this type are particle swarm optimization and differential evolution, which have been broadly used on real-valued optimization problems.

Brief outlines and pseudocodes were provided for each algorithm. In the following chapters, the exposed metaheuristics will be incorporated into specialized algorithm portfolio frameworks, aiming to tackle challenging optimization problems. The decision on which algorithms shall be eventually incorporated into an algorithm portfolio depends on their complementary properties as well as on the distinctive characteristics of the problem at hand.

Chapter 2
Algorithm Portfolios

Instead of tackling an optimization problem through the application of a single solver, exploiting the algorithmic power of multiple methods can increase the probability to detect better solutions in shorter running time. Based on this simple idea, algorithm portfolios are defined as algorithmic schemes that harness a set of algorithms, which are typically sharing the available computation and hardware resources.

In this chapter, we present the basics on algorithm portfolios by exposing their integral components and inherent mechanisms, as well as related concepts. Moreover, we outline three essential design issues of the algorithm portfolio framework, which will be further analyzed in the forthcoming chapters. These include the selection of constituent algorithms, the allocation of the available computation resources, and topics regarding sequential and parallel application. Effectively addressing these challenges makes the difference between success or failure of the algorithm portfolio when hard optimization problems are confronted.

2.1 Basics of Algorithm Portfolios

Algorithm portfolios are algorithmic schemes that combine a set of algorithms into a joint framework [58, 68]. The term "algorithm portfolio" was initially introduced in [68], inspired by established investment strategies that suggest building a mixed portfolio of financial assets (e.g., stocks) to maximize profit and minimize risk.

Motivated by this established economic approach, in algorithm portfolios, a set of algorithms act as assets, sharing the available computation resources to address hard optimization problems [68]. Up to now, portfolios have demonstrated their potential in several application fields including graph pattern matching [28], inventory routing [137], berth allocation [164], lot sizing [142], facility location [34],

13
D. Souravlias et al., *Algorithm Portfolios*, SpringerBriefs in Optimization,
https://doi.org/10.1007/978-3-030-68514-0_2

combinatorics [143], and cryptography [144]. Moreover, deep neural networks have been recently utilized for the automatic construction of algorithm portfolios [91].

For presentation purposes, we consider the bound-constrained minimization problem:

$$\min_{\mathbf{x} \in \mathcal{X}} f(\mathbf{x}),$$

where $\mathcal{X}$ stands for a n-dimensional bounded search space. The type of $\mathcal{X}$ (e.g., discrete, continuous, binary) is unimportant as it is related only to the type of solvers that are used. Also, let

$$A = \{a_1, a_2, \dots, a_k\},$$

be a set of k solvers (solution methods) available for the given problem. The set A may include different algorithms, different parameter configurations of the same algorithm, or both.

The first step toward the design of an algorithm portfolio is concerned with the algorithms that shall constitute it, also called the constituent algorithms. To this end, an algorithm selection procedure is initially applied to identify the M solution methods that will be included in the portfolio. Obviously, $M \leqslant k$, and the outcome of the algorithm selection phase is a portfolio AP defined as

$$AP = \{a_{i_1}, a_{i_2}, \dots, a_{i_M}\} \subseteq A.$$

In order to facilitate presentation, the indices $\{i_1, i_2, \dots, i_M\}$ will be henceforth denoted as $\{1, 2, \dots, M\}$, without loss of generality.

Therefore, an algorithm portfolio AP that consists of M constituent algorithms is henceforth denoted as a set

$$AP = \{a_1, a_2, \dots, a_M\}.$$

If different algorithms are included in AP, we have a heterogeneous portfolio. On the other hand, if multiple copies of a specific algorithm are used, adopting either the same or different parameter configurations, we obtain a so-called homogeneous portfolio.

The constituent algorithms may be of different types, e.g., evolutionary algorithms [5, 118], trajectory-based metaheuristics [140, 164], or SAT solvers [89, 167]. The present book is primarily devoted to algorithm portfolios that consist of either population-based or trajectory-based metaheuristics, such as the ones presented in Chap. 1. Note that population-based approaches are typically distinguished for their exploration capabilities, while trajectory-based approaches focus on local neighborhoods, thereby achieving enhanced exploitation performance.

The rationale behind algorithm portfolios lies in the fact that the best algorithm for a given problem is frequently a priori unknown. Therefore, blending together into a portfolio the algorithmic capability of (preferably) diverse metaheuristics is expected to increase the probability of finding a satisfactory solution. At the same time, it minimizes the risk that accompanies the selection of a single algorithm in terms of solution-quality deviation.

During the run of an algorithm portfolio, its constituent algorithms may communicate and exchange information with each other. This is the case of interactive algorithm portfolios. Alternatively, they can independently and interchangeably run in the case of non-interactive portfolio. Interactive portfolios are concerned with several design choices including the type and size of the exchanged information as well as the type of communication, i.e., synchronous or asynchronous. In the synchronous case, communication frequency needs to be determined. In the asynchronous case, the conditions that trigger communication shall be defined.

Overall, interactive portfolios promote collaborative optimization strategies where different algorithms cooperate, habitually achieving superior performance than stand-alone algorithms [118, 142]. Nevertheless, the decision on triggering interaction within a portfolio shall take into consideration possible complementarity properties of its constituent algorithms, as well as the inherent characteristics of the problem at hand.

An algorithm portfolio typically employs a resource allocation plan that determines how the available computation budget, i.e., execution time or function evaluations, is shared among its constituent algorithms. Putting it formally, let $T_{\max}$ be the available computation budget. Each constituent algorithm $a_i \in AP$, $i = 1, 2, \ldots, M$ is assigned a fraction, T_i, of the available budget, ensuring that

$$T_{\max} = \sum_{i=1}^{M} T_i.$$

Hence, an algorithm portfolio follows a resource allocation plan, $\mathcal{T}$, that corresponds to a set of assigned budgets defined as

$$\mathcal{T} = \{T_1, T_2, \ldots, T_M\}.$$

The constituent algorithms of the portfolio may sequentially and interchangeably run on a single processing unit or concurrently run on multiple processing units. In the first case, each execution of the constituent algorithm a_i consumes a predetermined fraction of its assigned budget T_i, before the next algorithm occupies the processing unit in turn. Therefore, the per algorithm assigned budget is not consumed at once, but rather in multiple subdivided parts also called episodes or batches.

Alternatively, the portfolio may exploit a parallel computation environment where the constituent algorithms share the available hardware resources. More formally, let

$$P = \{p_1, p_2, \ldots, p_C\},$$

be a set of C processing units devoted to the portfolio. Assuming this setting, the portfolio may employ a common master-slave parallel model where the master node focuses on the coordination of the portfolio's processes as well as the necessary bookkeeping, and the slave nodes host the constituent algorithms.

In the best case, the number of processing units is adequate to host all the constituent algorithms, each one running on its own slave node, as well as the master node:

$$C \geqslant M + 1.$$

If this is not possible, more than one constituent algorithms are assigned to each slave node, and they are executed on the same processing unit.

2.2   Design Challenges

The design of algorithm portfolios involves a number of challenges that shall be confronted by the practitioner [141]. In particular, the following aspects need to be addressed:

(a) Selection of the constituent algorithms.
(b) Allocation of the computation budget.
(c) Parallel vs sequential implementation.

The first aspect, i.e., the selection of the constituent algorithms that will be included in the portfolio, is relevant to the so-called algorithm selection problem originally described by Rice [131]. In the portfolio context, this selection is typically performed offline, i.e., prior to the application of the portfolio, taking into account the performance profiles of the algorithms available for the studied problem.

Performance profiles are hardly a priori known. Thus, the user shall consider a preprocessing phase where the algorithms are applied to the problem at hand until adequate information on their performance is collected. Such information may include the solution quality achieved by each available algorithm across different configurations and various computation budgets. Chapter 3 of the present book is exclusively devoted to this portfolio design challenge.

The second aspect is concerned with the allocation of the computation budget among the constituent algorithms of the portfolio. For this purpose, the portfolio typically uses a resource allocation plan defined either offline, namely, prior to the application of the portfolio, or online during its execution. Offline approaches construct fixed allocation plans based on diverse performance information. This information is gathered through individual application of each constituent algorithm on the studied problem. An example of such information is the average amount of computational budget that shall be consumed by an algorithm to achieve a specific solution quality. On the other hand, online approaches use the available historical data for their initiation. Then, they dynamically modify the initial resource allocation plan according to the performance of each algorithm during the optimization procedure. More detailed information on resource allocation in algorithm portfolios is presented in Chap. 4.

The last aspect refers to the sequential and parallel application of a portfolio framework. In the sequential case, the constituent algorithms are executed on a single processing unit in a round-robin manner. To this end, the portfolio employs a dedicated scheduling mechanism that defines which algorithm is applied next [172]. On the other hand, a parallel algorithm portfolio involves algorithms that share the available hardware resources, e.g., the processing units [140]. Such settings promote interaction among the constituent algorithms, thereby enhancing the overall performance of the parallel algorithm portfolio [118, 142]. Moreover, different parallel models can be used to boost the efficiency of a portfolio framework [5]. An overview of sequential and parallel implementation issues of algorithm portfolios is provided in Chap. 5.

Summarizing, the design of algorithm portfolios hardly qualifies as a trivial task. Achieving an effective design and implementation requires all the aforementioned aspects to be addressed point by point while considering (i) the special characteristics of the problem at hand, (ii) the properties of the constituent algorithms, and (iii) the practitioner's preferences and experience. It is the authors' belief that the design of a successful algorithm portfolios is a highly empirical task that involves the thorough analysis of its building blocks (constituent algorithms, resource allocation mechanisms) under a variety of parameter configurations and problem characteristics.

2.3   Recent Developments

In the following paragraphs, a number of recent research studies on algorithm portfolios are outlined. These studies appeared during the writing of the present book, which is indicative of the dynamic of the specific research field. Although they are not further analyzed in the remainder of this book, they explore novel concepts that can form the basis for promising future developments in algorithm portfolios.

An approach for the automatic construction of parallel algorithm portfolios is introduced in [90]. Different from similar works in the literature, the proposed approach is not based on the assumption that a given training set can sufficiently represent the considered problem cases. In practice, such an assumption is usually invalid for two main reasons. First, the training set may consist of a limited number of instances that are insufficient to cover all possible problem cases. Second, the training instances may be outdated or biased, thereby failing to satisfactorily capture the properties of the considered problem cases. In order to mitigate this issue, the generation of additional instances appears to be a suitable strategy. However, creating good training data in practice is far from a straightforward task.

The proposed approach in [90], called the generative adversarial solver trainer, aims to simultaneously generate additional training instances and construct a parallel algorithm portfolio. For this purpose, an idea similar to the way that generative adversarial networks work is adopted. Specifically, instance generation and portfolio construction constitute the main players of a typical adversarial game. Instance generation is focused on the generation of hard problem instances that cannot be effectively tackled by the current portfolio.

On the other hand, the portfolio construction is dedicated to the selection of a new solver, whose addition in the portfolio would enable tackling the new challenging instances more effectively. The competitive dynamics of the considered game eventually result in a parallel algorithm portfolio capable of effectively solving a higher number of problem instances, thereby leading in a more generalized solution method for the problem at hand.

The portfolio construction procedure results in a portfolio of M solvers that run in parallel and independently on each of the considered problem instances until a specific termination criterion is satisfied [90]. Different termination criteria may be applied, typically depending on the considered problem domain and the performance measure of interest. For example, whenever a decision problem (e.g., SAT problem) is under consideration, the portfolio is terminated as soon as one of its constituent algorithms provides an answer (either satisfiable or not). In this case, the required running time of the portfolio to solve a problem instance corresponds to the running time consumed by the portfolio's best constituent algorithm to solve the specific instance.

On the contrary, whenever an optimization problem is tackled (e.g., a TSP problem), a particular performance measure may act as the necessary termination criterion. If solution quality is the considered performance measure, then the portfolio terminates its run as soon as one of its constituent algorithms detects a solution of specific quality, e.g., within a predefined distance from the optimum.

The goal of the instance generation procedure is to create additional training instances [90]. Such instances shall ideally have the following properties:

(a) Constitute challenging tasks for the constituent algorithms of the portfolio.
(b) Cover target cases that are not already covered by the existing training instances.

Given a training set, I, a new instance s_{new} is created using a base instance $s \in I$ and a number of instances, also called reference instances, selected uniformly and randomly from $I \setminus \{s\}$. The new instance s_{new} is then generated by modifying s through random perturbation, along with the incorporation of components from the reference instances. Applying this procedure to each instance $s \in I$ results in the generation of a new instance set, I_{new}.

New instances in I_{new} differ significantly from the existing instances in I. Nevertheless, instances from both sets share some common characteristics. This is a highly desirable outcome because the new set will not be restricted in instances similar to that of I. In turn, this leads to better exploration of the instance space, promoting the construction of more generalized portfolios.

Another recent story of success for algorithm portfolios is their application on maritime logistics presented in [164]. More specifically, the proposed portfolio is used to solve large berth allocation problems under limited execution time budgets.

Berth allocation is one of the most important tasks in port terminals. It is concerned with generating schedules for the assignment of berths to incoming vessels subject to ready times and vessel size constraints. Solving this scheduling problem at the strategic level requires a vast number of simulations that account for different aspects including time horizons of months or years, high number of vessels, as well as different ship traffic scenarios that involve time uncertainties.

In practice, berth allocation problems appear in diverse variants. Hence, using a single solver to effectively solve each variant is an infeasible approach. This emphasizes a sheer need for the development of algorithmic schemes, such as algorithm portfolios, which allow the execution of multiple solvers, especially when large-scale problems are confronted under limited computation budgets [164].

Given a set of training instances, I', and a set of candidate algorithms, A, the goal is to select which algorithms will be eventually inserted in the portfolio. Let $x_a \in \{0, 1\}$ be a binary variable defined as

$$x_a = \begin{cases} 1, \text{ if algorithm } a \in A \text{ is included in the portfolio,} \\ 0, \text{ otherwise.} \end{cases}$$

The algorithm portfolio is formed according to the following integer linear program [164]:

$$\min Cost, \tag{2.1}$$

$$\sum_{a \in A} y(a, I, T_{\max})\, x_a \geqslant 1, \qquad \forall I \in I' \tag{2.2}$$

$$\sum_{a \in A} t(a, I, T_{\max})\, x_a \leqslant Cost, \qquad \forall I \in I', \tag{2.3}$$

where $y(a, I, T_{\max})$ is equal to 1 if algorithm a provides the best solution for instance I, given $T_{\max}$, and 0 otherwise. The term $t(a, I, T_{\max})$ stands for the running time required for algorithm a to solve instance I given $T_{\max}$. Detailed information on how to compute the value of this term can be found in [164]. The variable $Cost$ refers to the overall time cost required to run the portfolio. Therefore, in the above formulation, Eqs. (2.1) and (2.3) minimize the maximum time cost of applying the portfolio on any instance, while Eq. (2.2) ensures that each instance can be solved by at least one algorithm of the portfolio within the time budget $T_{\max}$.

The proposed portfolios combined different types of algorithms, including greedy algorithms, hill climbers (i.e., simple trajectory-based methods), as well as different variants of established metaheuristics such as greedy randomized adaptive search procedure (GRASP) and iterated local search. Initially, each algorithm was individually evaluated on randomly generated datasets. In order to compare solutions detected by different algorithms, the authors considered two quantitative measures. More specifically, they used the number of wins, i.e., the number of instances for which an algorithm provides the best solution, along with the number of unique wins, i.e., the number of instances for which an algorithm detected the best solution uniquely.

Then, the considered algorithms and test instances were used to construct algorithm portfolios by solving the integer linear programming model presented in Eqs. (2.1)–(2.3) through an exact solver. It is important to note that the construction of algorithm portfolios was achieved by applying an evolution in running time approach that considered different running time budgets. Overall, the proposed approach consistently identified suitable constituent algorithms for the portfolio, thereby demonstrating its effectiveness especially on large instances of the berth allocation problem [164].

Following a different path, the first study that considers the development of algorithm portfolios consisting of surrogate-assisted evolutionary algorithms is presented in [158]. Surrogate-assisted evolutionary algorithms are established solvers for computationally intensive optimization problems, especially the ones met in real-world applications. Even though a large number of surrogate-assisted evolutionary algorithms have been proposed in the literature, there is no approach suitable to address every problem equally well. Moreover, identifying the best algorithm for an unknown (computationally intensive) optimization problem is in practice a tedious task.

Motivated by these observations, the authors in [158] proposed two algorithm portfolio frameworks, called parallel individual-based surrogate-assisted evolutionary algorithm (Par-IBSAEA) and upper confidence bound individual-based surrogate-assisted evolutionary algorithm (UCB-IBSAEA), appropriate to address expensive optimization problems. Despite the operational differences in these frameworks, a limited budget of function evaluations was considered for benchmarking of both portfolios.

The Par-IBSAEA framework draws inspiration from established algorithm portfolios that simultaneously apply population-based algorithms, such as the ones proposed in [118]. However, instead of embedding typical evolutionary algorithms into the proposed framework, the algorithmic power of surrogate-assisted evolutionary algorithms was exploited. Specifically, the portfolio includes a number of different surrogate-assisted evolutionary algorithms, running in parallel without any interaction among them (i.e., independently).

Moreover, a database of samples for the construction of a surrogate model is assumed to be shared by all algorithms. At each iteration, an algorithm evaluates one or more solutions, which are then used to update the database of samples. The portfolio terminates its execution as soon as the total budget of function evaluations is consumed by its algorithms.

The UCB-IBSAEA framework is modeled as a multi-armed bandit problem [48] where the surrogate-assisted evolutionary algorithm is regarded as the arm, while the quality of solutions is used to quantitatively assess the action reward. In order to define reward, the solution's objective function value is linearly scaled with respect to the bounds of the problem. However, as the confronted problems are typically black boxes, it is not possible to obtain the actual bounds. Therefore, one idea is to estimate the bounds by considering the evaluated individuals that are stored in a database of samples (which are again used to compute the surrogates). At each iteration, the framework selects the best algorithm over a list of candidates, according to its bound and reward information.

The application of the selected algorithm results in a solution that is evaluated and then added in the database of samples. This is followed by an update step where the bounds and reward information of each algorithm of the portfolio are recomputed. Finally, the portfolio terminates its run when a predefined budget of function evaluations is consumed.

2.4 Synopsis

Algorithm portfolios are algorithmic schemes that harness multiple algorithms into a unified framework. In this chapter, we covered the basics on algorithm portfolios by providing an overview that illustrates their building blocks and analyzed related concepts. Moreover, we outlined key design aspects that need to be addressed for a successful algorithm portfolio implementation: the selection of its constituent algorithms, the allocation of the available computation resources, and topics regarding its sequential and parallel application.

These design aspects are further discussed in the forthcoming chapters in order to expose the reader to their individual roles and comprehend their impact as part of the algorithm portfolio framework.

Chapter 3
Selection of Constituent Algorithms

Algorithm portfolios were primarily introduced as computational models that prevent from shortcomings produced by the selection of inappropriate algorithms in solving hard optimization problems. The concurrent or interchangeable application of all the constituent algorithms of the portfolio aims at exploiting their individual advantages and suppressing their weaknesses through complementarity. Despite the inherent tolerance of algorithm portfolios on their constituent algorithms selection, it is reasonable to consider that careful selection can boost their performance.

The present chapter is devoted to this problem, which is also equivalent to the well-known per instance algorithm selection problem. In this context, a review of the prominent algorithm selection methods is offered. They are distinguished into two main categories with respect to (i) the use of machine learning methods or (ii) the use of statistical selection strategies borrowed from evolutionary computation. It shall be noted that all these methods are applied offline, i.e., before the actual application of the algorithm portfolio, consuming fractions of the available computational resources. Nevertheless, careful selection of the constituent algorithms can promote effectiveness and efficiency of the portfolio in hard optimization problems.

3.1 Algorithm Selection Problem

Given an optimization problem, the selection of the best-performing algorithm over a number of available solvers constitutes an intriguing task. The so-called algorithm selection problem was originally introduced in [131], and it has been a central research topic over the past four decades. Typically, algorithm selection is considered over a set of problem instances under consideration.

More specifically, given a predetermined set of algorithms, S; a set of problem instances, I; and a cost function

$$\mathcal{F} : S \times I \rightarrow \mathbb{R},$$

the widely known per instance algorithm selection problem is concerned with the detection of a mapping

$$\mathcal{G} : I \rightarrow S,$$

such that the cost function $\mathcal{F}$ is minimized over all the considered problem instances. Thus, it essentially solves the general problem

$$\min_{\mathcal{G}} \sum_{i \in I} \mathcal{F}\left(\mathcal{G}(i), i\right). \tag{3.1}$$

A similar problem arises in algorithm portfolios regarding the selection of their constituent algorithms [7, 151]. More specifically, given a predefined set of algorithms, S; a set of problem instances, I; and a cost function $\mathcal{F}$, the goal is to detect a subset of algorithms $\tilde{S} \subset S$ that minimizes $\mathcal{F}$ over all problem instances.

The best subset of algorithms eventually compounds the algorithm portfolio AP:

$$AP \overset{\text{def}}{=} \arg\min_{\tilde{S} \subset S} \mathcal{F}\left(\tilde{S}, I\right). \tag{3.2}$$

The cost function $\mathcal{F}$ shall be an aggregate measure that quantifies the performance of the entire portfolio, taking into account the individual achievements of each constituent algorithm over all problem instances. For example, the average objective function value over all constituent algorithms can be used for this purpose.

The constituent algorithms selection problem of Eq. (3.2) is essentially equivalent to the problem of Eq. (3.1) where, instead of mapping algorithms to problem instances, the goal is to specify the portfolio's algorithms based on the aggregate performance measure.

3.2   Algorithm Selection Methods

The per instance algorithm selection problem has been the main challenge in various research studies [75, 81, 105]. The available methods that can be adopted in algorithm portfolios are applied offline, i.e., before running the algorithm portfolio. Typically, the selection among candidate solvers is based on their performance profiles. As such profiles are frequently unavailable, a preprocessing phase of intense experimentation is required, where a group of best-performing algorithms is identified and subsequently admitted in the portfolio. This procedure may need significant computational resources depending on the number of candidate algorithms, problem hardness, and experimental intensity.

Prevalent methods are divided into two main categories, based on their underlying selection approaches. The first category comprises methods that rely on machine learning [87, 88, 167], while the second category includes methods based on statistical selection [7, 104, 151, 173]. Methods of both types are reviewed in the following paragraphs.

3.2.1 Feature-Based Selection

Machine learning approaches typically employ feature-based mechanisms [71], properly designed to identify special features in families of related problems. These features concern the running time required by the candidate algorithms to solve the problems. Thus, they are used to build and train learning models that predict the performance of candidate algorithms on the specific problem type.

This is achieved by determining a representative training set of problems and collecting the corresponding running-time data of each candidate algorithm. Although the selection of the training set is nontrivial, properly designed experimentation can provide relatively accurate models that offer useful performance predictions on unseen problems of the specific type. Consequently, in view of a new problem instance, the predicted best-performing algorithm is selected.

In [87], the per instance algorithm selection problem is addressed via a knowledge-based method. The method is dedicated to the selection of special problem features that can provide indications regarding the running time required by the algorithms. This is achieved by using a set of problem instances that is initially generated according to a predefined distribution. The problems are tackled one by one by the available algorithms, and the time required to solve each one is recorded. This information is subsequently exploited by a statistical regression technique that learns a real-valued function of the preselected features.

The algorithm portfolios constructed with the knowledge-based method execute the following steps:

(a) The method is employed to train a model for each constituent algorithm.
(b) Feature values of the problem instance at hand are computed, and predictions of the algorithms' running times are obtained.
(c) Eventually, the algorithm that is predicted to require the lowest running time is selected and applied.

The proposed algorithm portfolio was tested on the combinatorial auctions problem, and it was shown to outperform its three constituent algorithms, individually, even though two of them were much slower than the best algorithm.

In a different approach [167], the special automated SATzilla framework for building per instance algorithm portfolios on SAT problems was extended. Given a set of problem instances and a group of solvers, SATzilla selects the constituent algorithms of the portfolio through the optimization of a proper objective function. This function is based on a performance measure oriented in running time. The automated framework is enhanced with hierarchical hardness models that exhibit increased accuracy in predicting the running time of each algorithm, regardless of the satisfiability of the specific problem instance. Thus, for each algorithm, a hierarchical hardness model is constructed to predict running time of the algorithm on each problem instance, using the problem's features.

Additionally, this model introduces local search algorithms into the portfolio and automates the pre-solver selection phase of the SATzilla framework, which was manually performed. These modifications resulted in boosting the effectiveness of SATzilla, rendering it capable of optimizing more complex functions.

In [88], various generic strategies to select and run a parallel portfolio of n SAT solvers are proposed. In order for these methods to be scalable on the number of solvers and processing units, the selection phase shall be polynomial in the size of the portfolio. Additionally, several assumptions are made regarding the construction of the parallel portfolio:

(a) Only deterministic algorithms are considered in the portfolio.
(b) A different algorithm per processing unit is selected.
(c) The algorithms run in parallel without any interaction.
(d) No further problem-specific knowledge is available, i.e., there is no information on the different types of the SAT instances.

In the parallel portfolio case, the selection mechanisms such as the ones employed by the majority of sequential selection methods (e.g., SATzilla [167]) shall be properly modified.

First, the algorithms of the portfolio are assigned scores for each problem instance through a scoring function. This function is computed according to several prominent feature-based selection methods, including specialized nearest-neighborhood techniques, clustering methods, and regression approaches.

Second, the algorithms are ranked according to their assigned scores. The n best-performing algorithms per instance are selected and, subsequently, run, each one on a different processing unit. Experimental results using four processing units revealed that the best approaches achieved up to 12-fold average speedup against the best individual solver, over a number of hard combinatorial problems [88].

3.2.2   Statistical Selection

An alternative to machine learning selection is the use of statistical techniques applied in evolutionary computation [7, 104, 151, 173]. These methods employ algorithms' scoring based on statistical comparisons according to their running times.

In [7], a general selection scheme is proposed. Given a number of available algorithms and a set of problem instances, it determines offline the group of best-performing solvers that will compound the portfolio. This is achieved by introducing diverse selection strategies (selectors), including a racing approach that is typically used for metaheuristics configuration.

The performance of the proposed selectors in [7] was evaluated by applying the produced algorithm portfolios on preventive maintenance scheduling problems particularly generated for benchmarking. Comparisons among six different selectors were conducted. The experimental analysis revealed that two selectors ("delete instance" and "racing") exhibited promising performance. Both policies achieved the best trade-off between solution quality and running time compared to other selectors.

The selector "delete instance" takes as input a group of available algorithms along with a list of problem instances and iteratively applies each algorithm on a randomly selected problem instance. If it is successfully solved (i.e., the quality of the acquired solution is above a user-defined threshold), it is removed from the list, and the corresponding algorithm is added in the portfolio. In case that none of the available methods manages to solve the specific problem instance, the algorithm that achieves the best solution is considered as the winner. Eventually, the portfolio comprises all algorithms that achieved to solve at least one problem instance.

The selector "racing" uses the well-known F-Race algorithm [29] to select a predefined number of solvers from an initial set of candidates. Following this approach, worst-performing algorithms are gradually discarded from the initial set, based on statistical comparisons of their online performance. The elimination of some algorithms spares computational resources for the rest. Thus, it allows their application on additional problem instances, enhancing the accuracy of their performance estimation. Eventually, the algorithms that survive this procedure to the end compose the algorithm portfolio.

In [151], a portfolio of population-based metaheuristics is introduced to address real-valued optimization problems. An automated technique is proposed to select a subset of the available algorithms in order to build the portfolio. The technique records the performance of each candidate algorithm, individually. Then, it exploits this information to build a per algorithm specialized matrix, also called the estimated performance matrix. For the construction of this matrix, the algorithm is applied k_1 times on each one of the available k_2 problem instances, consuming a specific fraction of the computational resources that are allocated to the whole portfolio.

Thus, for each algorithm, a matrix of size $k_1 \times k_2$ is generated containing the best solution value computed at each run of the algorithm on the corresponding problem instance. A cost function for each subset of the initial set of algorithms is then calculated by using those performance matrices. The algorithms that belong to the subset with the lowest cost value are selected as the constituent algorithms of the portfolio.

The algorithm selection phase is succeeded by an initialization step, where the populations of the constituent algorithms are initialized. Specifically, each population primarily admits the best solution it has detected during the selection phase. Its remaining individuals are randomly selected from the k_1 sub-populations that were generated in the matrix construction phase. Finally, the portfolio is used to solve (sequentially) each one of the considered problems until the allocated computation budget is consumed.

Another generic method for automatic construction of algorithm portfolios is proposed in [173]. Given an initial set of population-based metaheuristics, a ranking mechanism is employed to rank the algorithms according to their efficiency in solving a set of test problems coming from an established test suite. Each algorithm is applied on each problem until it attains a solution of predefined quality, and the required computation budget (function evaluations) is recorded. If the specified quality cannot be reached, the algorithm terminates after a maximum number of function evaluations. This outcome is also stored for the specific algorithm-problem pair.

Based on the acquired information, pairwise statistical comparisons are conducted between the algorithms by applying the widely used Friedman test [50]. The assessment is based on the efficiency of the algorithms in terms of the required average number of function evaluations over all runs for each problem, individually. Therefore, each metaheuristic is assigned a per problem rank between 1 and the maximum number of available algorithms, where lower ranking indicates better performance. If two algorithms achieve the same rank for a particular problem, their average performance over the considered test problems is used as a tiebreaker.

Eventually, the portfolio admits only the best metaheuristic, i.e., the one with the lowest average rank over all problems. An alternative variant admits also one additional algorithm in the portfolio. In this case, the two approaches shall exhibit performance complementarity. This means that the additional algorithm shall outperform the best-performing one at least on some of the considered problems. A covariance matrix is constructed using the ranking scores of the algorithms, in order to identify the complementary metaheuristic.

A voting-based method for building algorithm portfolios with complementary evolutionary algorithms on continuous black-box optimization problems is proposed in [104]. The so-called ICARUS method models the portfolio construction as an election procedure. The method consists of a preliminary stage and an iterative elimination stage. The preliminary stage is dedicated to the generation of an initial subset of algorithms, which includes all approaches that achieve the best performance for at least one of the problems. Then, in the iterative elimination stage, the winners of the election, i.e., the algorithms that are preferred by the majority of the problems, are identified using a typical voting system.

More specifically, each problem is considered as a voter that selects (votes for) a number of algorithms from a set of candidate solvers. Thus, each problem is associated with a preference list, in the form of a linear ordering of the considered algorithms, according to their performance scores. The scores quantify the algorithms' efficiency and accuracy in terms of their average expected running times as well as their percentage of successfully solved problems, respectively. Eventually, the selected algorithms are inserted into portfolio. Experimental evidence in [104] suggests that portfolios constructed using ICARUS can outperform manually constructed portfolios on an established continuous optimization test suite.

Further comprehensive studies on algorithm selection methods for combinatorial and continuous optimization problems are offered in [81] and [105], respectively. A recent survey of state-of-the-art algorithm selection methods for discrete and continuous problems as well as an overview of promising problem-specific features can be found in [75].

3.3 Synopsis

The construction of algorithm portfolios is a problem equivalent to the well-known algorithm selection problem. In the algorithm portfolio context, this problem refers to the selection of a group of appropriate high-quality algorithms with respect to predefined performance measures over a set of test problems. The chapter outlined prominent algorithm selection methods that have been used for the construction of algorithm portfolios in the relevant literature. Such methods are either working offline or online, with the first type outnumbering the second. The exposed methods were distinguished into two main categories, namely, feature-based methods that employ machine learning techniques and statistical methods that exploit statistical methodologies to compare the performance of the candidate algorithms.

Regardless of the selection method, the decision on which solvers shall become parts of an algorithm portfolio can pose biases on performance. The magnitude of this impact is highly related also to the specific problem under consideration.

Chapter 4
Allocation of Computation Resources

Resource allocation is a critical problem in algorithm portfolios. It refers to the schemes adopted for distributing the available computation resources (running time, function evaluations, or processing units) among the constituent algorithms. The computation resources can be assigned before the application of the portfolio (offline) or during its run (online). In the first case, the total computation budget allocated to each constituent algorithm is fixed and prespecified, while, in the latter, it is dynamically assigned in fractions during its execution.

Offline methods use historical or preprocessing data to identify the most promising algorithms. Thus, they proportionally assign computation resources based on their known quality, neglecting their performance in the current run. On the other hand, online methods adopt allocation schemes based on performance measures computed in the current run. Therefore, they can dynamically capture the relevant performance of the algorithms and dynamically assign the available computation budget, favoring the best-performing ones. The present chapter analyzes algorithm portfolios with online resource allocation approaches, which constitute the core methodologies of the present book.

4.1 General Resource Allocation Problem

The resource allocation problem is concerned with the distribution of the available computation budget among the constituent algorithms of the portfolio. Let T_{max} be the total computation budget, in terms of execution time or function evaluations, assigned to the whole portfolio prior to its application on the problem at hand. Each constituent algorithm, denoted as a_i, $i \in \{1, 2, \ldots, M\}$, receives a fraction T_i of the total budget, with

© The Author(s), under exclusive license to Springer Nature Switzerland AG 2021
D. Souravlias et al., *Algorithm Portfolios*, SpringerBriefs in Optimization,
https://doi.org/10.1007/978-3-030-68514-0_4

$$T_{\max} = \sum_{i=1}^{M} T_i. \tag{4.1}$$

A resource allocation plan, denoted as $\mathcal{T}$, consists of a set of allocated budgets:

$$\mathcal{T} = \{T_1, T_2, \ldots, T_M\}.$$

This plan can be determined following either an offline or an online approach.

Offline approaches are applied prior to the application of the portfolio, allocating computation resources according to each algorithm's performance profile on the problem at hand. This allocation remains fixed, i.e., it is unaltered during the optimization procedure. Hence, it neglects performance fluctuations observed during the application of the algorithms.

A naive assignment of this type is the allocation of equal amount of resources to each algorithm:

$$T_i = \frac{T_{\max}}{M}, \quad \forall i.$$

Naturally, this choice is prone to assignments that spend resources on weak algorithms, reducing the overall efficiency of the portfolio. A different type of offline assignment uses historical data such as the probability of successfully solving problems of the same type as the considered one. For example, if $\xi_i \in [0, 1]$ is the computed probability of successfully solving similar problems using a specific amount of resources for each algorithm a_i, then the assigned budgets T_i can be set as

$$T_i = \text{round}\left(\frac{\xi_i T_{\max}}{\Xi}\right), \quad \forall i,$$

where

$$\Xi = \sum_{i=1}^{M} \xi_i,$$

using appropriate rounding to the nearest integer through round( ) to ensure that Eq. (4.1) holds. Such data-based approaches, although improving the aforementioned naive scheme, have a number of disadvantages:

(a) The existing data may not be completely representative for the current experimental setting due to possible differences in algorithm configuration or to the specific parametrization of the problem at hand.
(b) Small changes in problem formulation or algorithm configuration would require the repetition of the preprocessing phase anew.

(c) They disregard the fact that each algorithm has its own special dynamic in different phases of the optimization procedure. This means that its effectiveness can vary at the beginning or at the end of the optimization procedure, depending on its exploration/exploitation capabilities.

Therefore, offline assignments are more suitable on problems of identical type, using performance data collected in preprocessing phases such as the algorithm selection presented in the previous chapter.

The weaknesses of offline methods leave ground for the development of alternative approaches, namely, online resource allocation methods [49, 140, 142]. Online approaches are based on resource allocation mechanisms that dynamically adjust the allocation plan according to the algorithms' current performance profile on the specific problem. This is achieved by collecting performance data during the specific run and using appropriate performance measures to assign quality-related grades to each constituent algorithm of the portfolio [49, 140]. Recently proposed online approaches of this type are presented in the following paragraphs.

4.2   Online Resource Allocation

Three recent algorithm portfolio models that employ online resource allocation mechanisms are analyzed below. For each portfolio, a detailed description of the underlying allocation mechanism is provided along with pseudocodes of its internal procedures. All the presented mechanisms share two common characteristics:

(a) They strive to assign computation resources on the basis of performance, such that best-performing algorithms are rewarded higher computation budgets than the rest.
(b) The budget allocation follows a greedy strategy. This means that a locally best allocation plan is specified, rather than aiming at a possible globally optimal one (if it exists).

It shall be noted that proper resource allocation is itself a nontrivial optimization task. Its outcome increases the probability but cannot guarantee the detection of optimal solutions.

4.2.1   Algorithm Portfolio with Trading-Based Resource Allocation

Stock trading is a complex system where stocks (the goods) are transferred among investors (the players) in exchange for money (resources). Various concepts from stock trading models offered inspiration for the development of the algorithm portfolio framework introduced in [142].

In that framework, the portfolio's constituent algorithms act as investors that invest on elite (high-quality) solutions (the analogue for stocks), using portions of their allocated time as the currency. The core of this approach is a trading-based mechanism where fractions of the assigned computation budget of the algorithms are transferred among them in exchange of good solutions. This way, best-performing algorithms sell their findings to the rest while earning higher computation budgets that extend their running time. Note that the use of this mechanism does not modify the total execution time $T_{\max}$ that is pre-allocated to the whole portfolio.

The portfolio proposed in [142] assumes M metaheuristics that run in parallel, exchanging information during the optimization procedure. Although parallelization is not a prerequisite, it offers significant gain in time complexity. For this purpose, a typical master-slave parallelization model is employed, where each algorithm runs on a devoted slave node. A master node is also used for bookkeeping, i.e., storing and pricing elite solutions, as well as forwarding them to the constituent algorithms whenever needed.

The total running time $T_{\max}$ is initially assigned to the portfolio and distributed equally among the constituent algorithms. In the market analogue, this means that all investors start with the same amount of money. Then, each algorithm a_i divides its own budget, $T_{\max}^{(i)}$, into execution time, $T_{\text{exec}}^{(i)}$, and investment time:

$$T_{\text{inv}}^{(i)} = \lambda \, T_{\max}^{(i)},$$

where the parameter $\lambda \in (0, 1)$ configures each algorithm's investment policy. Obviously, higher λ values correspond to riskier investment policies. The assigned execution time is consumed by the algorithm itself, while investment time is used by the algorithm to purchase elite solutions from the rest, whenever its search stagnates.

The master node performs two essential operations:

(a) It retains a pool of M elite solutions $\mathbf{x}_i^*$, one for each constituent algorithm.
(b) It applies a pricing procedure to the stored solutions.

Pricing quantifies the solutions' value, i.e., the investment time that an algorithm shall pay in order to buy a solution. To this end, the considered solutions are sorted in descending order with respect to their objective function values. Then, the price C_i of the stored solution $\mathbf{x}_i^*$ is computed taking into account its place ρ_i in the sorted order, as follows:

$$C_i = \frac{\rho_i \, BC}{M}, \qquad (4.2)$$

where

$$BC = \theta \, T_{\text{inv}}, \qquad \theta \in [0, 1].$$

The parameter BC is also called the base cost, and θ is a parameter that controls the algorithm's purchasing power, having a tangible effect on its elitism. Higher θ values result in smaller number of elite solutions that can be bought by the algorithm during the optimization procedure. This is in line with stock markets, where best-performing stocks are more expensive and, thus, given a specific money budget, and only a limited number can be bought.

After their initialization, the constituent algorithms run on the problem at hand. Whenever algorithm a_i, running on the i-th slave node, improves its own best solution, the new solution is sent to the master node, replacing its previous best solution $\mathbf{x}_i^*$. In the case that an algorithm does not improve its own best solution for some time, T_{imp}, a purchase procedure is activated.

In the purchase procedure, the algorithm attempts to purchase from the master node one of the stored elite solutions (excluding its own) that can be exploited in its subsequent execution. For example, a population-based metaheuristic can introduce the purchased solution in its population, while a trajectory-based method can use this solution (or a mildly perturbed one) as the starting point for a new trajectory.

In order to assist the buyer algorithm a_i to make a profitable purchase, a solution selection mechanism is triggered. This mechanism chooses an elite solution $\mathbf{x}_j^*$, $i \neq j$, based on an established investment performance index, called the return on investment (ROI):

Algorithm 6 Algorithm portfolio with trading-based resource allocation: pseudocode of the master node

Input: Computation budget T_{max}; execution time T_{exec}; investment time T_{inv}; parameter $\theta \in [0, 1]$; number of algorithms M

Output: Best detected solution $\mathbf{x}^*$

1: $S \leftarrow$ **initialize** $-$ **archive**(M)
2: **while** (active slave nodes exist) **do**
3:      **get** $-$ **message**(i, s) /* receive message s from the i-th slave node */
4:      **if** ($s =$"new solution") **then**
5:          **get** $-$ **solution**$(i, \mathbf{x}_i^*, f_i^*)$
6:          $S \leftarrow$ **replace** $-$ **solution**$(S, \mathbf{x}_i^*, f_i^*)$
7:      **else if** ($s =$"buy solution") **then**
8:          $S \leftarrow$ **sort** $-$ **decreasing**(S, f^*)
9:          **for** ($j = 1 \ldots, M, j \neq i$) **do**
10:            $C_j \leftarrow$ **pricing**$(\mathbf{x}_j^*, f_j^*)$
11:            $ROI_j \leftarrow (f_i^* - f_j^*)/C_j$
12:          **end for**
13:          $[\mathbf{x}_k^*, f_k^*] \leftarrow$ **buy** $-$ **solution** $\left(i, T_{\text{inv}}^{(i)}, S, ROI\right)$
14:          **send** $-$ **solution**$(i, \mathbf{x}_k^*, f_k^*, C_k)$
15:          **increase** $-$ **seller** $-$ **time**(k, C_k)
16:      **end if**
17:      $\mathbf{x}^* \leftarrow$ **best** $-$ **solution**(S)
18: **end while**
19: **return** $\mathbf{x}^*$

$$ROI_j = \frac{f_i^* - f_j^*}{C_j}, \quad j \in \{1, 2, \ldots, M\}, \tag{4.3}$$

where f_i^* stands for the objective value of the buyer's best solution, f_j^* denotes the objective value of the seller's elite solution, and C_j is the price assigned to $\mathbf{x}_j^*$ by Eq. (4.2). The buyer algorithm a_i decides to buy the solution $\mathbf{x}_j^*$ that fulfills the following criteria:

(a) It maximizes the ROI index.
(b) It has better objective function value than the algorithm's own solution stored in the elite set.
(c) The algorithm can afford the solution, i.e., its investment budget is adequate:

$$T_{inv}^{(i)} > C_j.$$

If the algorithm does not have adequate budget to buy the overall best solution, it buys the first affordable solution with the highest ROI value that improves its own solution.

The purchase is completed in two steps. In the first step, a copy of the new solution along with its objective function value is forwarded from the master node to the corresponding slave node occupied by a_i. In the second step, the seller algorithm, i.e., the algorithm that discovered the sold solution $\mathbf{x}_j^*$, is rewarded with an amount of time equal to C_j. More specifically, the buyer algorithm decreases its own investment budget by C_j, whereas the seller algorithm prolongs its own execution time by C_j:

Algorithm 7 Algorithm portfolio with trading-based resource allocation: pseudocode of the i-th slave node

---

Input: Algorithm a_i; execution time $T_{exec}^{(i)}$; investment time $T_{inv}^{(i)}$; no improvement time $T_{noimp}^{(i)}$

 1: **initialize** $-$ **node**(i)
 2: **while** $\left(T_{exec}^{(i)} > 0 \right)$ **do**
 3:      **apply** $-$ **algorithm**(a_i)
 4:      **if** (new best solution $\mathbf{x}_i^*$ detected) **then**
 5:          $s \leftarrow$ "new solution"
 6:          **send** $-$ **message**(i, s)
 7:          **send** $-$ **solution**($i, \mathbf{x}_i^*, f_i^*$)
 8:      **end if**
 9:      **if** (no improvement for T_{noimp}) AND $\left(T_{inv}^{(i)} > 0 \right)$ **then**
10:          $s \leftarrow$ "buy solution"
11:          **send** $-$ **message**(i, s)
12:          **get** $-$ **solution**($i, \mathbf{y}, f_y, C_y$)
13:          $T_{inv}^{(i)} \leftarrow T_{inv}^{(i)} - C_y$
14:          **incorporate** $-$ **solution**($a_i, \mathbf{y}$)
15:      **end if**
16: **end while**

$$T_{\text{inv}}^{(i)} \leftarrow T_{\text{inv}}^{(i)} - C_j, \qquad T_{\text{exec}}^{(j)} \leftarrow T_{\text{exec}}^{(j)} + C_j.$$

The procedure continues iteratively until the total computation budget is exceeded. Note that the total budget of the portfolio is unaffected by the procedure but remains fixed to T_{max}. Also, the execution times of the algorithms are not simultaneously changed, but rather asynchronously, as soon as one of the algorithms stagnates. Algorithms 6 and 7 report the pseudocodes for the master and the slave nodes, respectively, based on the implementations in [142]. The message passing interface (MPI) [61] software is a straightforward choice for the implementation of the parallel model. A demonstration of the trading-based model on a difficult combinatorial optimization problem is postponed until Chap. 6.

4.2.2 Algorithm Portfolio Based on Performance Forecasting

A different type of algorithm portfolio was proposed in [140], where a forecasting mechanism is used to predict the performance of the constituent algorithms. The predictions are based on three established models for time series forecasting, namely, the simple exponential smoothing [94, 96, 134], the linear exponential smoothing [96], and the simple moving average [69, 96].

The main goal is to use the computed predictions to synchronously allocate computation resources among the constituent algorithms according to their forecasted performance. In this way, better-performing algorithms gradually extend their running times, while shorter budgets are assigned to the rest, without being eliminated.

Let AP denote the portfolio consisting of M algorithms:

$$AP = \{a_1, a_2, \ldots, a_M\}.$$

Let also $C \geqslant M$ be the number of available processing units in terms of physical cores or processing threads. Additionally, let T_{max} stand for the total computation budget allocated to the portfolio. In [140], the maximum number of function evaluations was considered for this purpose. However, running time can be alternatively used with proper modification. The execution of the portfolio is terminated as soon as the available function evaluations are fully consumed by its constituent algorithms.

According to the specific portfolio model, the available function evaluations are distributed among the constituent algorithms in batches. The number of batches, b_{max}, is defined by the user. Then, the computation budget allocated to the constituent algorithms per batch is given as

$$T_{\text{batch}} = \frac{T_{\text{max}}}{b_{\text{max}}}. \tag{4.4}$$

The use of batches plays a crucial role in the coordination of the portfolio operations. In particular, the end of each batch constitutes a synchronization point where all constituent algorithms temporarily pause their execution, enabling the forecasting mechanism to be subsequently applied.

For this reason, using function evaluations to define a batch is deemed as a suitable option, especially if the processing units belong to a heterogeneous computation environment. Indeed, algorithms executed on faster processing units consume more function evaluations than the rest within a specific time interval. Thus, the straightforward use of running time as computation budget would introduce performance bias, rendering the comparisons between the algorithms unfair.

The computation budget T_{batch} assigned to each batch is equally divided among the C processing units. Thus, each processor is allocated a budget T_{proc} defined as

$$T_{\text{proc}} = \frac{T_{\text{batch}}}{C}. \tag{4.5}$$

The main decision lies in distributing the constituent algorithms to the available processing units, i.e., determine how many of the C processing units will be occupied by each constituent algorithm.

Let p_b^i denote the number of processing units on which algorithm a_i is executed, during batch b. The sum of the occupied processing units shall be

$$C = \sum_{i=1}^{M} p_b^i, \tag{4.6}$$

for each batch $b \in \{1, 2, \ldots, b_{\text{max}}\}$. The computation budget assigned to batch b is completely consumed by the constituent algorithms before the assignment of the next batch. Thus, algorithms that occupy faster processing units may need to pause until all processing units finish their execution.

The best objective value achieved by each algorithm since the beginning of the optimization procedure is recorded. This information is used by the forecasting mechanism to predict the forthcoming performance of each constituent algorithm. Finally, the predictions are used to determine the exact number of processing units that each algorithm will occupy at the next batch, $b + 1$.

Let f_b^i denote the overall best objective function value achieved by algorithm a_i at batch b, regardless of the number of processing units it occupies. Without loss of generality, it is assumed that f_b^i takes only non-negative values. Also, let $\hat{f}_b^i$ stand for the predicted objective function value of algorithm a_i at batch b, which is computed with any of the available forecasting models. The set

$$H_b^i = \left\{ f_1^i, \hat{f}_2^i, f_2^i, \hat{f}_3^i, f_3^i, \ldots, \hat{f}_b^i, f_b^i \right\},$$

$$i \in \{1, 2, \ldots, M\}, \qquad b \in \{1, 2, \ldots, b_{\text{max}}\},$$

Algorithm 8 Forecasting-based algorithm portfolio

Input: Algorithms $a_1, \ldots, a_M$; number of processing units C; number of batches $b_{\max}$; computation budget $T_{\max}$

Output: Best detected solution $\mathbf{x}^*$

1: $T_{\text{proc}} \leftarrow T_{\max}/(C\, b_{\max}),\ b \leftarrow 0$
2: /* assignment for 1st batch */
3: **for** $(i = 1 \ldots M)$ **do**
4:     $\eta_{b+1}^i \leftarrow 1/M$
5:     $p_{b+1}^i \leftarrow$ **allocate** $\left(\eta_{b+1}^i, C\right)$
6:     $H_b^i \leftarrow \emptyset$
7: **end for**
8: **assign-cpus**$(p_{b+1}^i, i = 1 \ldots M)$
9: /* loop on the number of batches */
10: **for** $(b = 1 \ldots b_{\max})$ **do**
11:     *[NODE i]*: **execute** $-$ **algorithm** $\left(a_i, p_b^i, T_{\text{proc}}\right)$ /* parallel execution on slave nodes */
12:     **for** $(i = 1 \ldots M)$ **do**
13:         **update** $-$ **best** $\left(f_b^i\right)$
14:         $H_b^i \leftarrow H_{b-1}^i \cup \{f_b^i\}$
15:         $\hat{f}_{b+1}^i \leftarrow$ **forecaster** $\left(H_b^i\right)$
16:         $H_b^i \leftarrow H_b^i \cup \{\hat{f}_{b+1}^i\}$
17:         $p_{b+1}^i \leftarrow$ **allocate** $\left(\eta_{b+1}^i, C\right)$
18:     **end for**
19:     **assign-cpus**$(p_{b+1}^i, i = 1 \ldots M)$
20: **end for**
21: **return** $\mathbf{x}^*$

---

contains the actual and the predicted function values achieved by algorithm a_i from batch 1 up to batch b. Additional information may be incorporated in this set based on the selected forecasting model (e.g., the history length of the moving average model). Using this set, the function value $\hat{f}_{b+1}^i$ for algorithm a_i in the next batch $b + 1$ is predicted by applying the selected forecasting model:

$$\hat{f}_{b+1}^i = \textbf{forecaster}\left(H_b^i\right).$$

For example, using the order-k moving average forecasting model, the predicted value would be

$$\hat{f}_{b+1}^i = \frac{1}{k} \sum_{j=0}^{k-1} f_{b-j}^i.$$

Naturally, any forecaster that employs the information stored in H_b^i can be used for the same purpose.

Then, the percentage of processing units, η_{b+1}^i, occupied by algorithm a_i in the next batch $b + 1$ is defined as

$$\eta^i_{b+1} = \frac{\frac{1}{\hat{f}^i_{b+1}}}{\sum_{j=1}^{M} \frac{1}{\hat{f}^j_{b+1}}}, \quad \forall i, b. \tag{4.7}$$

Note that normalization is used to ensure that η^i_{b+1} takes values in $[0, 1]$.

Finally, the actual number of processing units, p^i_{b+1}, occupied by algorithm a_i in batch $b + 1$ constitutes a fraction of the C available processing units:

$$p^i_{b+1} = \mathbf{allocate}\left(\eta^i_{b+1}, C\right), \quad \forall i, b, \tag{4.8}$$

where **allocate**() stands for the resource allocation mechanism. The employed mechanism must ensure that Eq. (4.6) always holds, i.e., the number of processing units occupied by all algorithms does not exceed C. A straightforward way to implement the allocation mechanism is to set p^i_{b+1} to the nearest integer of the decimal number $\eta^i_{b+1} C$. However, this approach may lead to a total number different than C as a result of the rounding process, thereby violating Eq. (4.6).

This issue can be addressed in various ways. An idea adopted in [140] is to make a preliminary assignment of the algorithms to the C processing units as follows:

$$\left\lfloor \eta^i_{b+1} C \right\rfloor, \quad \forall i, b. \tag{4.9}$$

This scheme always assigns less than C processing units. The remaining ones are occupied by an algorithm in a round-robin assignment, based on their η^i_{b+1} values. Therefore, the algorithm with the best (highest) fraction value is assigned the first spare unit, the second best algorithm is assigned the second spare unit, and this procedure continues until each available processing unit is occupied by an algorithm.

The forecasting-based portfolio in [140] is based on a parallel implementation according to a typical master-slave model [142, 143], where the master node is responsible for bookkeeping and coordination of the slave nodes, whereas the slave nodes are dedicated to the execution of the constituent algorithms. The pseudocode given in Algorithm 8 provides the main steps of the presented algorithm portfolio. All lines of the provided pseudocode that refer to bookkeeping, forecasting, and resource allocation operations are sequentially executed on the master node, with the exception of line 11 that refers to the execution of the corresponding algorithm on each slave node. An application of the forecasting-based portfolio on a combinatorial optimization problem is provided in Chap. 6.

4.2.3 Adaptive Online Time Allocation Framework

A general online algorithm portfolio framework that allocates computation time to constituent algorithms was introduced in [49]. Given a set of algorithms, A, and a set of problems, R, the main goal is to allocate higher running time budgets to the most promising algorithms. These include the best-performing algorithms according to an adaptive model that is built using their search experience.

To this end, an online time allocation mechanism was developed, which automatically gathers information on the past performance of the algorithms and allocates the available computation resources, accordingly. The devised mechanism consists of distinct steps. At step k, the mechanism solves problem $r(k) \in R$ using algorithm $a(k) \in A$ and discrete computation time $t(k)$. When $t(k)$ is consumed, the algorithm pauses its execution and reports a vector $d(k) \in \mathbb{R}^D$. This vector contains information on the success of $a(k)$ converging on $r(k)$, as well as on the improvement achieved by applying $a(k)$ to $r(k)$.

Putting it more formally, for all steps k, the pair

$$(a(k), t(k)) = f(r(k), H(k), P), \tag{4.10}$$

is a function of the cumulative historical experience

$$H(k) = \{i, r(i), a(i), t(i), d(i) : 0 < i < k\},$$

and an initial bias P, which is typically defined as a probability distribution on A that is modified online to reflect each algorithm's performance on the considered problems. The function f can be determined through different learning algorithms, such as artificial neural networks trained on recent historical values of $H(k)$ or support vector machines used to predict the vector $d(k + 1)$ by employing information included in different combinations of $a(k + 1)$ and $t(k + 1)$.

In [49], extrapolating linear models that receive as input past performance data of the used algorithms are proposed for this purpose. Let $r(k) = r$ for all k, i.e., there is only a single problem, and $A = \{a_1, a_2, \ldots, a_M\}$ be a set of M genetic algorithms adopting different parametrizations. Also let

$$P(k) = \{p_i(k) = Pr\{a_i\} \text{ at step } k, \ i = 1, \ldots, M\},$$

be the bias defined as a probability distribution over A. The distribution is determined by applying a function f_P to a set of values $U(k) = \{u_i(k), i = 1, \ldots, M\}$ that aims to quantify the performance of each algorithm at step k through a predetermined function f_u.

Also, the vector $d(k)$ contains the average fitness value of the individuals in the population of the algorithm $a(k)$. Thus, the history

$$H_i(k) = \{(t(k), d(k)); \ a(k) = a_i\},$$

of a_i is a table that holds the average function value for each time step k. The computation time is then divided into small time slots of duration ΔT, and a sequence of pairs $(a(k), t(k)) = (a_i, p_i \Delta T)$ is specified where the time slots are allocated to the algorithms according to the considered bias. The pseudocode of the adaptive time allocation procedure is presented in Algorithm 9.

Regarding the function f_u, for each step k and algorithm a_i, the quantity $u_i(k)$ is computed as follows:

$$u_i(k) = \frac{1}{T_{i,sol} - T_i(k)}, \tag{4.11}$$

where $T_{i,sol}$ is the time required for algorithm a_i to reach the target fitness value and $T_i(k)$ is defined as the time consumed by algorithm a_i up to time step k:

$$T_i(k) = \sum_{a(k)=a_i} t(k).$$

For the computation of $u_i(k)$ defined in Eq. (4.11), $T_{i,sol}$ is initially estimated by a linear regression model that receives as input a shifting window of the c most recent values in $H_i(k)$. More specifically, $T_{i,sol}$ becomes the time at which the linear model predicts that algorithm a_i will reach the target fitness value, rounded up to the nearest multiple of the function evaluations required to run one generation of algorithm a_i. Note that the considered number of function evaluations is equal to the algorithm's population size.

In the employed linear model, c shall be sufficiently small to enable rapid reaction to any trend changes of the average fitness values. Also, it shall be large enough to make up for possible noise in the average fitness values. Additionally, experimental evidence suggests against the use of the same window size for all the algorithms. For this reason, adaptive window size is employed in the adaptive online allocation procedure, namely, the following three window sizes:

$$c_i, \quad 2c_i, \quad \max(c_i, 2c_i), \quad i \in \{1, 2 \dots, M\},$$

which define three different linear models. At step k, the new vector $d(k)$ is compared with the values predicted by the three models. The window size that corresponds to the model with the lowest prediction error becomes the current c_i.

Regarding the function f_P, simple normalization is employed as follows:

$$p_i(k) = \frac{u_i(k)}{\sum_{j=1}^{M} u_j(k)}, \tag{4.12}$$

with $p_i(0)$ being inversely proportional to population size z_i for all a_i. Additionally, the value of u_i for $k = 0$ is set as

---

Algorithm 9 Pseudocode of the adaptive online time allocation framework

---

Input: Problem $r(k) = r$ for all k; set of algorithms $A = \{a_1, a_2, \ldots, a_M\}$; bias $P(k) = \{p_i(k) = Pr\{a_i\}$ at step $k, i = 1, \ldots, M\}$

Output: Allocation of time slots to algorithms

 1: **initialization**$(\Delta T, u_i(0), i = 1 \ldots M)$
 2: $k \leftarrow 1$
 3: **while** (problem r not solved) **do**
 4:     $P_A(k) \leftarrow f_P(U_A(k))$
 5:     **for** $(i = 1 \ldots m)$ **do**
 6:         $a(k) \leftarrow a_i, t(k) \leftarrow p_i \Delta T$
 7:         **run-pair**$(a(k), t(k))$
 8:         $u_i(k + 1) \leftarrow f_u(H_i(k))$
 9:         $k = k + 1$
10:     **end for**
11: **end while**
12: **return** $(a(k), t(k)) = (a_i, p_i \Delta T)$ for all $i \in \{1, \ldots, M\}$

$$u_i(0) = \frac{\zeta}{z_i},$$

where ζ is a constant.

The adaptive allocation procedure above aims at algorithm independence. This is because the used algorithms are treated as black boxes, assuming that no information on their structural characteristics and internal operations is concealed. Additionally, each algorithm a_i individually runs, and the same holds for updating u_i.

Interaction among the algorithms occurs indirectly through their competition for execution time during the normalization step for the computation of f_P in Eq. (4.12). Algorithm-related knowledge is restricted on specifying the set of algorithms A, the state information d, and the bias P. A comprehensive presentation of the procedure is offered in [49].

4.3 Synopsis

We presented three online resource allocation schemes that are used within algorithm portfolio frameworks. The underlying models aim at distributing the available computation resources among the constituent algorithms in an online fashion. In all cases, the allocation mechanism is based on the constituent algorithms' performance during the search procedure. Two schemes are based on estimations of future performance, while another scheme is based on a novel model where an algorithm interacts with the rest upon search stagnation. The chapter underlined the key concepts of each model and offered general descriptions of the involved procedures as well as pseudocodes that can be used as the basis for effective implementations.

Chapter 5
Sequential and Parallel Models

Algorithm portfolios are implemented either sequentially or in parallel. In the sequential case, the constituent algorithms interchangeably run on a single processing unit, consuming fixed fractions of their allocated computation resources at each turn. The parallel case harnesses the computational efficiency of multiple processing units, as presented in the resource allocation models of the previous chapter. Each constituent algorithm runs either on a devoted processing unit, consuming all its allocated resources in batches, or it occupies multiple processing units, each one consuming a fixed portion of the total amount of resources provided to the portfolio. The present chapter reviews state-of-the-art portfolio frameworks for each model type, along with their distinctive characteristics and algorithmic peculiarities.

5.1 Sequential Models

Sequential algorithm portfolios comprise models where the constituent algorithms interchangeably run on a single processing unit [66, 118, 172]. This was the first type of algorithm portfolios reported in the relevant literature [68]. Although it was focused on a specific algorithm type, it served as proof of concept for motivating further development of algorithm portfolio models. In the following paragraphs, we review three essential sequential portfolio models, based on metaheuristics for tackling real-valued optimization problems. The specific approaches were chosen on the ground of popularity [118, 172] and novelty [66]. Thus, they can be used as the starting point for the development of more sophisticated new approaches.

The first of the considered algorithm portfolios was proposed in [118] for solving numerical optimization problems within limited time budgets. The portfolio combines m population-based search algorithms, each one consuming a fraction of the total time budget while running on a single processing unit. During the portfolio's

© The Author(s), under exclusive license to Springer Nature Switzerland AG 2021
D. Souravlias et al., *Algorithm Portfolios*, SpringerBriefs in Optimization,
https://doi.org/10.1007/978-3-030-68514-0_5

run, information sharing among the constituent algorithms is implemented in terms of a solution migration procedure.

This procedure involves the migration of individuals among the populations. For this purpose, it employs two user-defined integer parameters:

(a) Migration size ($s > 0$): it defines the number of migrant solutions.
(b) Migration period ($p > 0$): it defines the number of generations between two successive migrations.

Moreover, the approach in [118] considers time budget in terms of function evaluations that are allocated to the portfolio and distributed among its constituent algorithms. In particular, a joint population is considered and divided into disjoint sub-populations, one for each constituent algorithm.

Within this framework, all the constituent algorithms are used to cooperatively evolve the entire population. Every p generations, migration of individuals takes place among sub-populations of all algorithms. More specifically, for each algorithm a_i, the sub-populations of the remaining algorithms are combined, and the best s individuals in terms of objective function value are identified. Then, these individuals are used to replace the worst s individuals of the algorithm a_i.

An extensive evaluation of the considered portfolio was based on 11 portfolio instances produced by combinations of four state-of-the-art population-based algorithms. An established benchmark suite consisting of 27 objective functions was used in the simulation experiments. The analysis of the results showed the superiority of the portfolios compared to their constituent algorithms, individually. More specifically, 7 out of 11 portfolio instances outperformed their constituent algorithms with respect to solution quality. Also, 7 out of 11 portfolio instances detected solutions of equal or better quality than the G-CMA-ES algorithm [26], which was superior against all the constituent algorithms. Moreover, the conducted analysis revealed that complementarity of the constituent algorithms is a key factor for building portfolios of supreme quality.

Another algorithm portfolio that incorporates a number of evolutionary algorithms to tackle single-objective optimization problems was introduced in [172]. The main research question investigated in that study was as follows:

> Given a portfolio of algorithms, which algorithm should I choose to be executed next?

To this end, the proposed algorithm portfolio is equipped with a parameter-less prediction mechanism that determines which algorithm should be applied next, based on its former accomplishments. The selected algorithm runs for a single generation after which, the performance of each constituent algorithm is predicted anew. This way, the constituent algorithms are interchangeably executed on a single processing unit by consuming only a part of the available computational resources.

In the core of the prediction mechanism lies a novel performance evaluation measure that considers a family of convergence curves for each algorithm [172]. Each curve represents the evolution of the best solution value over the iterations (generations) of the algorithms. More specifically, let the data points

$$C(l) = \{(t - l, f_{t-l}), (t - l + 1, f_{t-l+1}), \ldots, (t, f_t)\},$$

define a sub-curve that includes all data points from the $(t - l)$-th to the t-th generation, where f_t is the best solution value at generation t and l is the history size. A simple linear regression model is applied to fit each sub-curve $C(l)$ to a linear line, which is subsequently used to predict l function values for the $(t + 1)$-th iteration. These function values are then employed to fit a bootstrap probability distribution that, in turn, generates the final predicted function value. Eventually, the algorithm that achieves the best predicted function value is selected to run for the next generation.

Experimental results using an established benchmark suite showed that the proposed method outperformed several existing algorithm portfolios as well as simpler approaches [172]. Moreover, comparisons between the portfolio and its constituent algorithms were conducted, with the proposed portfolio ranked second best. This can be attributed to the fact that a significant part of the computational resources was spent to determine the best algorithm within the portfolio to be executed next. In these comparisons, there were also some cases where the portfolio was identified as the best approach, leading to the conclusion that the synergy among its algorithms can have a positive impact on solution quality.

A third sequential portfolio that can accommodate any optimization algorithm and requires minor implementation effort was introduced in [66]. The main issues addressed in that work were:

(a) Which constituent algorithm shall run next.
(b) When an algorithm shall pause its run.

The study extended the typical algorithm portfolio framework by incorporating an algorithm selection approach and a dedicated termination mechanism.

The algorithm selection approach is based on an offline training procedure that treats the algorithm selection problem as a classification task. Given a set of algorithms and a set of problem instances, each instance is classified to a specific algorithm label. In order to initially specify the labels, the algorithms are executed on each problem instance, and the best-performing one (i.e., the algorithm that achieved the best solution) is used to label the specific instance. Next, each algorithm is applied to a problem instance for a specific number of function evaluations, and a set of exploratory landscape analysis features is computed. The labels and the extracted features form a per instance data matrix that is subsequently given as input to the classification task. As the accuracy of the classification result is sensitive on the sample size of the employed features, multiple classification models are exploited by the algorithm selection approach.

The termination mechanism automatically stops the running optimization algorithm if it fails to achieve significant improvement within a specific time frame. Given the population size N of a population-based algorithm, and a number of function evaluations T, the mechanism periodically decides whether the algorithm shall be stopped every $2T$ evaluations, i.e., every $(2T)/N$ generations. At each generation, the best solution value is recorded. Also, the best function values of the first and last T/N iterations, respectively, are stored.

At each detection period, an established statistical test is applied to compare the statistical significance of the difference between the two sets. If no statistical significance is encountered, then the best objective value has not been improved for T evaluations, and thus, the algorithm is terminated.

5.2 Parallel Models

One of the most important developments in computing has been the introduction of a variety of parallel models and the development of solution methods that exploit the computation power of parallel computers to successfully tackle challenging optimization problems [8, 46, 113–115]. In this direction, parallel portfolios have been introduced to take full advantage of the capabilities of multiple processing units [58]. For portfolios, parallelism is considered as a mean of boosting (i) effectiveness, in terms of solution quality, and (ii) efficiency, in terms of running time [5, 140, 142, 143]. The most popular parallelization approach is the master-slave model [5, 142], where the master node is the coordinator, while the algorithms run on the slave nodes. Enabling different search algorithms to probe the search space in parallel offers the evident benefit of reduced running time when compared to the sequential approach. In addition, solution quality can be improved as the concurrent execution of different solvers amplifies the exploration capabilities of the portfolio.

A parallel algorithm portfolio was introduced in [5]. It comprises three well-studied population-based metaheuristics, namely, artificial bee colony [72], differential evolution [145], and particle swarm optimization [37]. Two parallel approaches were proposed, based on the coarse-grained parallel model and the master-slave model, both implemented using the MPI framework.

The first approach employs the coarse-grained parallelization model, where the constituent algorithms are concurrently executed on each available node, having a sub-population assigned to each algorithm. Migration is performed at certain time intervals to communicate each algorithm's individual search experience to the rest of the constituent algorithms. Given that the portfolio consists of m algorithms, the employed migration strategy replaces the worst individual of each algorithm's sub-population with the best individual detected by the other $m - 1$ constituent algorithms.

The second approach builds the portfolio according to a two-level parallelization scheme. The upper level exploits a coarse-grained parallel model that enables the independent run of the constituent algorithms on each node. The lower level makes use of a master-slave model, where the master node distributes the individuals among the slave nodes, and each slave node is dedicated to computing the fitness value of a single individual. Upon completion, all fitness values are forwarded from the slaves nodes to the master node, which is responsible for performing the necessary selection/update actions.

The time efficiency of the produced parallel implementations under various parameter settings was studied in [5]. Continuous test problems from an established benchmark suite, as well as time series forecasting problems, were used for benchmarking purpose. Experimental results showed that the proposed parallel portfolios achieved more robust performance while also frequently outperforming their sequential counterparts.

In [174], a different parallel algorithm portfolio model was proposed. This model is oriented toward constrained satisfaction problems, accommodating case-based reasoning, greedy algorithms, as well as a set of heuristics. Before forming parallel portfolios, sequential portfolios are defined through an offline construction technique that learns the performance of the constituent algorithms on a set of training problems and uses this learning outcome to address the entire set of testing problems. Typically, this technique relies on a feature-extraction mechanism applied on a set of training problem instances, along with a performance matrix that is used to store the time required by each constituent algorithm to solve a training instance. The result of the feature-extraction mechanism is fed to a sequential portfolio constructor focused on generating a per problem schedule that determines which algorithm will tackle the problem at each time interval. The portfolio constructor in [174] was properly modified to exploit similarities among problems in order to provide more accurate estimations on the running time required for a specific training problem.

In order to produce portfolios suited for parallel processing, an integer programming formulation was introduced in the same study. The main objective was to find the optimal time allocation of the constituent algorithms on the available processors, such that the running time offered to each problem by all processors is less than a specific value, and the portfolio achieves to successfully solve the maximum possible number of problems.

For this purpose, the sequential portfolio constructor is properly extended in the parallel case by adding different heuristic methods dedicated to assigning algorithms to the available processors. Experimental results revealed that the developed portfolio constructor created statistically superior parallel portfolios than the ones built by naive parallel versions of well-known portfolio constructors.

In [89], the automatic configuration of parallel portfolios from a single sequential solver or a set of sequential solvers was studied. In order to construct parallel portfolios from a single solver, three different approaches were proposed, and their corresponding performance was investigated:

(a) A technique called Global that simultaneously configures all constituent algorithms of the parallel portfolio. Specifically, if the considered solver has l parameters, then a portfolio of k algorithms is treated as a single algorithm of $l * k$ parameters.

(b) A technique named as ParHydra that iteratively adds algorithms to the portfolio, specifying the configuration of one algorithm at each iteration, while the rest remain intact.

(c) A technique that initially forms clusters in a space of instance features and, then, combines the best configuration of each cluster into a parallel portfolio.

Experimental results showed that the ParHydra approach outperforms the rest of the approaches, achieving comparable performance to a state-of-the-art parallel solver.

In order to automatically construct parallel portfolios from a set of different solvers, the methods used for automatically configuring parallel portfolios from a single solver were properly extended. In particular, the extension requires the use of some additional parameters, namely, a choice parameter per constituent solver, which determines that the specific solver is selected. Also, it requires some conditional parameters, which guarantee that a certain configuration for the selected solver is chosen. Empirical results on SAT problems showed that parallel algorithm portfolios combining different complementary solvers can outperform portfolios based on a single SAT solver. It is important to note that this result can be accomplished with marginal modifications on the typical automatic construction techniques that work using a single solver.

Finally, a framework that hybridizes Global and ParHydra to configure parallel portfolios consisting of multiple sequential and parallel solvers was introduced. Unlike ParHydra, the proposed hybrid approach simultaneously configures b constituent solvers of the portfolio at each iteration, simultaneously. This choice introduces additional overhead, which is compensated by the enhanced ability of the hybrid approach to avoid local minima. Experimental results showed that combining the hybrid construction approach with expert knowledge leads to automatically built parallel portfolios that outperform state-of-the-art parallel SAT solvers.

Another master-slave parallelization model was proposed in [142, 143]. The portfolio consists of a master node that retains an archive of elite solutions detected by the constituent algorithms and a number of slave nodes devoted to the execution of the constituent algorithms. Each time an algorithm detects a better overall solution, it is forwarded to the master node where the incumbent elite solution of the same algorithm is replaced. The proposed parallel portfolio was demonstrated on a well-studied manufacturing problem [142] as well as on the detection of circulant weighing matrices problem [143]. Details on the specific mechanism were given in Chap. 4.

Another pure master-slave model for parallel algorithm portfolios was proposed in [140], where time series forecasting techniques are used to allocate the available computational resources among the constituent algorithms. The master node is dedicated to bookkeeping, forecasting, and resource allocation procedures, while the slave nodes are devoted to the constituent algorithms' execution. The efficiency of that scheme was demonstrated on a family of circulant weighing matrices problems. More details are presented in Chap. 4.

5.3 Synopsis

The concurrency of execution of the constituent algorithms provides an explicit distinction between sequential and parallel algorithm portfolios. In the sequential case, constituent algorithms are interchangeably executed on a single processing unit. On the other hand, parallel portfolios exploit the computational power of multiple processing units.

In this chapter, existing approaches of both types were reviewed, and their main characteristics were outlined. In general, information sharing among different constituent algorithms used by some of the presented approaches has proved to be beneficial on the performance of the portfolio. For this reason, it is highly suggested for future developments.

5.3 Synopsis

Chapter 6
Recent Applications

Algorithm portfolios have proved to be promising solvers capable of solving demanding optimization problems in diverse application fields. In this chapter, we present three recent applications of algorithm portfolios on problems from the fields of combinatorics and operations research.

The first application refers to the detection of circulant weighing matrices, a special type of matrices met in various applications including cryptography and coding theory.

The second application is about lot sizing planning in production systems with returns and remanufacturing. The main goal is finding the number of new items that need to be produced, and the fraction of returned items that need to be remanufactured, in a production environment where transformation of already used products into like-new ones is possible.

The third application tackles the problem of commodities transportation in humanitarian logistics, where a number of basic commodities such as water, food, and medicines shall be transported to areas affected by natural or man-made disasters. In this case, the main goal is the optimal management of available transportation resources in order to completely satisfy the demand of the basic commodities, taking into consideration hard constraints such as operational limitations of the transportation vehicles and limited capacity.

For each one of these challenges, we provide a general description of the problem and the employed mathematical model, along with presentation of the algorithm portfolios proposed in the relevant references. Application-related aspects are also analyzed, including the algorithms that constitute the portfolio, their interactions, and their proper parametrization. Compact presentation and discussion of the reported results verify that the proposed algorithm portfolios are capable of tackling the particular problems, outperforming their constituent algorithms as well as other solvers.

© The Author(s), under exclusive license to Springer Nature Switzerland AG 2021
D. Souravlias et al., *Algorithm Portfolios*, SpringerBriefs in Optimization,
https://doi.org/10.1007/978-3-030-68514-0_6

6.1   Application in Detection of Combinatorial Matrices

Combinatorial matrices constitute a class of square matrices that possess prescribed combinatorial properties. A special family of combinatorial matrices consists of circulant weighing matrices, and it has increasingly attracted attention of researchers for many years. Applications of this matrix type can be found in diverse scientific fields including coding theory, where they are used for the development of linear codes with properties of substantial quality [12], and quantum information processing, where they are employed to boost the performance of quantum algorithms [40]. Their significant practical impact is also witnessed in other fields such as statistical experimentation and optical multiplexing [82].

The existence of finite or infinite classes of circulant weighing matrices constitutes an essential research challenge in combinatorics. For this purpose, two complementary research directions appear in the relevant literature: theoretical approaches and computational developments. Theoretical approaches and, particularly, algebraic methodologies have been proposed to determine the necessary existence conditions for specific types of circulant weighing matrices [11, 13, 44, 45, 53].

In cases where algebraic approaches fall short, computational methods have been used. The application of computational methods requires the conversion of the matrix existence problem into an equivalent discrete optimization (usually minimization) problem [35, 38, 77–79] whose global optimizers correspond to the desirable matrices. Besides the existence problem, the classification of certain types of circulant weighing matrices has been also considered in the relevant literature [9, 14, 135].

6.1.1   Problem Formulation

A square $n \times n$ matrix $W = [w_{ij}]$ with entries

$$w_{ij} \in \{-1, 0, 1\}, \quad i, j \in \{1, 2, \ldots, n\},$$

is called a weighing matrix of order n and weight k^2 and denoted as

$$W(n, k^2),$$

if there exists a positive integer $k < n$ such that

$$W W^T = k^2 I_n, \tag{6.1}$$

where I_n stands for the identity matrix of size n and W^T denotes the transpose matrix of W. A circulant weighing matrix, denoted as

$$CW(n, k^2),$$

is a weighing matrix of order n and weight k^2 satisfying the property that, excluding its first row, each other row is a right cyclic shift of its preceding one. Thus, this type of matrix can be completely defined solely by its first row. Note that when $n = k^2$, then a $CW(n, n)$, for n being 1,2, or a multiple of 4, is a circulant Hadamard matrix.

Theoretical methodologies have been applied to detect circulant weighing matrices of various orders and weights [9, 14, 45, 146]. Aside from the theoretical approaches, computational optimization algorithms have been used to solve this problem [77, 78, 140, 143]. Prerequisite of their application is a preparatory phase in which the original problem is transformed into an equivalent permutation optimization problem. The outcome of this phase is a specific objective function whose global minimizers define desirable circulant weighing matrices, i.e., each global minimizer specifies the first row of one desirable matrix.

In order to define the objective function of the permutation optimization problem, the periodic autocorrelation function is used [77]. More specifically, let

$$T^n = \{(x_1, x_2, \ldots, x_n), \ x_i \in \{-1, 0, +1\} \text{ for all } i\},$$

be the set of all ternary sequences of length n with elements $x_i \in \{-1, 0, +1\}$. Also, let

$$\mathbf{x} \in T^n,$$

be such a ternary sequence that, along with its right cyclic shifts, defines a $CW(n, k^2)$ matrix. Then, the periodic autocorrelation function associated with the ternary sequence $\mathbf{x} \in T^n$ of length n is defined as

$$\mathrm{PAF}_{\mathbf{x}}(s) = \sum_{i=1}^{n} x_i \, x_{i+s}, \quad s = 0, 1, \ldots, n-1, \tag{6.2}$$

where $i + s$ is taken modulo n when $i + s > n$. An interesting class of matrices is obtained if we demand that

$$\mathrm{PAF}_{\mathbf{x}}(s) = 0, \quad s = 0, 1, \ldots, n-1. \tag{6.3}$$

All ternary sequences $\mathbf{x} \in T^n$ that satisfy the system of Eq. (6.3) will be henceforth called admissible sequences, and they possess the symmetry property

$$\mathrm{PAF}_{\mathbf{x}}(s) = \mathrm{PAF}_{\mathbf{x}}(n-s), \quad s = 0, 1, \ldots, n-1, \tag{6.4}$$

i.e., each ternary sequence $\mathbf{x} \in T^n$ has a symmetric sequence with equal PAF value. Thus, it is adequate to consider only half of the admissible ternary sequences for the solution of the system of Eq. (6.3). Moreover, it can be proved that the admissible sequences have exactly:

(a) $n - k^2$ components equal to 0,
(b) $k(k + 1)/2$ components equal to +1,
(c) $k(k - 1)/2$ components equal to −1.

To enable the reader to fully understand the form of circulant weighing matrices, we provide the examples of the matrices CW(7, 4) and CW(4, 4) as follows:

$$CW(7, 4) = \begin{pmatrix} -1 & 1 & 1 & 0 & 1 & 0 & 0 \\ 0 & -1 & 1 & 1 & 0 & 1 & 0 \\ 0 & 0 & -1 & 1 & 1 & 0 & 1 \\ 1 & 0 & 0 & -1 & 1 & 1 & 0 \\ 0 & 1 & 0 & 0 & -1 & 1 & 1 \\ 1 & 0 & 1 & 0 & 0 & -1 & 1 \\ 1 & 1 & 0 & 1 & 0 & 0 & -1 \end{pmatrix}, \quad CW(4, 4) = \begin{pmatrix} -1 & 1 & 1 & 1 \\ 1 & -1 & 1 & 1 \\ 1 & 1 & -1 & 1 \\ 1 & 1 & 1 & -1 \end{pmatrix}$$

It is easy to see that the first row of each one of the above matrices has an autocorrelation value of 0.

Based on all the aforementioned properties, the derived combinatorial optimization problem is eventually defined as

$$\min_{\mathbf{x} \in T_A^n} f(\mathbf{x}) = \sum_{s=1}^{\lceil \frac{n}{2} \rceil - 1} |PAF_{\mathbf{x}}(s)|, \tag{6.5}$$

where $T_A^n \subset T^n$ is the set of admissible sequences of length n . Obviously, each global minimizer of Eq. (6.5) corresponds to an admissible ternary sequence that defines a $CW(n, k^2)$ matrix. Note that the existence of circulant weighing matrices is not guaranteed for each length n and weight k^2 . A collection of the solved and open problems of this type can be found in [13].

Taking into consideration the fixed number of appearances of 0, +1, and −1 in the admissible sequences, the detection of a circulant weighing matrix becomes a permutation optimization problem, where the main goal is to find the optimal ordering of the three values in the sequence. Various approaches including metaheuristics have been extensively used for solving such problems [77, 78]. In general, the level of difficulty of the problem increases with the length of sequence n and the weight k^2 .

6.1.2 Known Existence Results

In this section, known existence results are outlined with respect to the fundamental question: For a given weight k, what are the possible values of n so that a matrix $CW(n, k^2)$ exists?

1. A $CW(n, 4)$ exists if and only if n is even ($\neq 2$) or $7|n$. This result was proved in [45].
2. A $CW(n, 9)$ exists if and only if $13|n$ or $24|n$. This result was proved by Strassler in [146] using a computer search. About a decade later, a theoretical proof of this result was provided in [9], using the concept of a proper circulant weighing matrix.
3. In [24], Arasu and Seberry use a product theorem and some known circulant weighing matrices to show the existence of a number of circulant weighing matrices. Algebraic techniques are used to rule out the existence of some circulant weighing matrices.
4. A $CW(n, 16)$ exists if and only if $n \geq 21$ and $14|n, 21|n, 31|n$. This result was proved in [106], [19]. However, the classification of $CW(n, 16)$ for odd n was initially done by R. M. Adin, L. Epstein, and Y. Strassler in [3].
5. In [20], the authors study circulant weighing matrices of weight 2^{2t} for some positive integer t and obtain new structural results.
6. If q is a prime power, then there exists $CW(q^2 + q + 1, q^2)$. This result is connected to the theory behind finite projective planes [62].
7. If q is a prime power, q odd, and i even, then there exists $CW(\frac{q^{i+1}-1}{q-1}, q^i)$.
8. If $q = 2^t$ and i is even, then there exists $CW(\frac{q^{i+1}-1}{q-1}, q^i)$.
9. If there exists $CW(n_1, k^2)$ and $CW(n_2, k^2)$ with $(n_1, n_2) = 1$, then there exist $CW(mn_1, k^2)$ for all positive integers m and two inequivalent $CW(n_1 n_2, k^2)$, $CW(n_1 n_2, k^4)$.
10. In [85], the authors prove that for every odd prime power q, there are at most finitely many proper circulant weighing matrices of weight q. In [86], the same authors prove that any such proper circulant weighing matrix of weight q has the order $v \leq 2^{q-1}$.
11. In [84], the authors show that a proper $CW(n, 25)$ exists if and only if $n \in \{31, 62, 124, 71, 142, 33\}$.
12. In [135], the authors classify all circulant weighing matrices whose order and weight are products of powers of 2 and 3. In particular, they show that proper $CW(n, 36)$ exists for all $n \equiv 0 \,(mod\ 48)$.

6.1.3 23 Years of Strassler's Table: 1997–2020

The PhD thesis of Yoseph Strassler [146] was a work of pivotal importance in the research on circulant weighing matrices. It contained the complete classification of circulant weighing matrices of weight 9. Moreover, it contained a table that summarized the state-of-the-art knowledge on the existence question for circulant weighing matrices $CW(n, k^2)$, $n \in [1, 200]$, $k \in [1, 10]$. This table became the driving force behind subsequent research on circulant weighing matrices in the past 23 years and has become known under the name of *Strassler's table*. Several researchers contributed minor and major updates in Strassler's table. These updates include new theoretical and computational results, on existence and non-existence questions. The methods being used in these updates range from algebraic number theory to character theory, metaheuristics, and beyond. Below we summarize some of these updates, and we note that 23 years after its inception, there are still open/undecided cases in Strassler's table.

1. The first example of $CW(71, 25)$ was given by Strassler in [147].
2. In [25], an algebraic method is used to construct $CW(33t, 25)$ for each positive integer t.
3. A first significant update of Strassler's table was given in [21]. Arasu and Ma propose a few structural theorems for circulant weighing matrices whose weight is the square of a prime number. Their results provide new schemes to search for circulant weighing matrices. They also establish the status of several previously open cases, i.e., they show non-existence of $CW(147, 49)$, $CW(125, 25)$, $CW(200, 25)$, $CW(55, 25)$, $CW(95, 25)$, $CW(133, 49)$, and $CW(195, 25)$.
4. The first major update of Strassler's table was given in [13]. More specifically, Arasu and Gutman resolved 52 open/undecided entries in Strassler's table. In all 52 cases, it turns out that circulant weighing matrices with these parameters do not exist.
5. Several examples of $CW(48, 36)$ were given in [80] using algorithm portfolios and extensive computation on multicore processors. Examples of $CW(48, 36)$ were first discovered by Arasu and Kotsireas in 2011, using two $CW(24, 9)$ with disjoint support (unpublished manuscript).
6. The first example of $CW(142, 100)$ was given in [18], using the concept of *disjoint support* and a variation of a theorem from [16]. Using two $CW(71, 25)$ with disjoint support, a $CW(142, 100)$ was constructed in [18].
7. In [23], the authors prove the non-existence of $CW(154, 36)$ and $CW(170, 64)$. This is an application of a theorem that established a necessary condition for existence of $CW(2n, k^2)$, assuming there is no $CW(n, k^2)$.
8. In [22], the authors prove the non-existence of $CW(110, 100)$.
9. In [170], the author proves the non-existence of $CW(117, 25)$, $CW(133, 25)$, $CW(152, 25)$, $CW(171, 25)$, $CW(148, 49)$, $CW(162, 49)$, and $CW(198, 49)$, using modular constraints and multipliers.
10. In [15], the authors prove the non-existence of $CW(88, 81)$ and $CW(99, 81)$ by applying purely counting methods for the former and algebraic methods for the latter.
11. The second major update on Strassler's table was given in [150]. In this study, Tan resolved 18 open/undecided entries in Strassler's table. She used techniques from algebraic number theory, namely, the field descent method and Weil numbers. In all 18 cases, it turns out that matrices with these parameters

do not exist. The same author proves in her PhD thesis the non-existence of $CW(60, 36)$ [149], which was the smallest long-standing open problem for several years.

12. The first examples of $CW(126, 64), CW(198, 100)$ were found computationally in [10] using the Douglas-Rachford algorithm. The reader is referred to [30] for a complete description of the Douglas-Rachford iteration scheme. While these circulant weighing matrices were listed as open in the most recent version of Strassler's table at the time, they can in fact be constructed via Theorem 2.2 presented in [16] as follows:

 (i) Take $CW(21, 16)$ (well known) and use $m = 3$ to get $CW(126, 64)$.
 (ii) Take $CW(33, 25)$ [25] and take $m = 3$.

13. In [17], the authors prove the non-existence of $CW(104, 81)$, $CW(110, 81)$, $CW(154, 81)$, $CW(130, 81)$, $CW(143, 81)$, and $CW(143, 36)$.

Even though there have been numerous research results on the (non-)existence of different circulant weighing matrices, the interest in tackling open cases that appear in Strassler's table remains current and vivid. To find out if a circulant weighing matrix of specific weight and order exists or not, the reader can access the website maintained by Dan Gordon [39].

6.1.4 Solving the Problem Through Metaheuristics

Metaheuristics have been successfully applied to address challenging circulant weighing matrices problems [77, 78, 140, 143]. Among them, state-of-the-art local search methods such as tabu search and variable neighborhood search, as well as global search methods such as particle swarm optimization and differential evolution, have offered promising results. Essential concepts of these metaheuristics are exposed in Chap. 1.

Tabu search and variable neighborhood search have been primarily designed as discrete optimization solvers. Thus, they can be directly applied to the derived combinatorial problem of Eq. (6.5). On the contrary, population-based metaheuristics such as particle swarm optimization and differential evolution primarily aim at continuous optimization problems. Therefore, their application on the studied combinatorial problem requires an appropriate decoding procedure for the translation of real-valued vectors into ternary permutation vectors.

A simple yet efficient technique for this translation is the smallest position value representation [152]. According to this scheme, an initial reference vector that includes the exact number of 0, $+1$, and -1 components is generated for the specific circulant weighing matrix problem. For example, the reference vector can contain all its 0 components in its first $n - k^2$ positions, then its $+1$ components in the following $k(k + 1)/2$ positions, and in the last $k(k - 1)/2$ positions its -1 components. Then, a n-dimensional real-valued vector can be considered as a vector of priority weights, where the i-th real-valued component defines the priority of appearance of the corresponding i-th reference component in the ternary sequence with, supposedly, higher real values denoting higher priority. Thus, the real-valued vector is sorted in descending order, and the corresponding components of the reference vector are placed according to the derived order, building a ternary sequence. Note that this approach can map an infinite number of real-valued vectors of the same order among their components to a specific ternary sequence. Thus, it basically defines classes among the real-valued vectors.

Another critical issue in the application of metaheuristics on the circulant weighing matrices problem is parallelization [143]. Parallel implementations primarily boost efficiency in terms of running time. Furthermore, they can also increase effectiveness, as several instances of an algorithm running in parallel can simultaneously explore diverse parts of the search space [6]. A typical parallelization framework for the specific problem is based on the standard master-slave model, where each slave node runs a copy of the same algorithm with identical or different parameter configuration. Interaction among the nodes occurs in terms of a migration mechanism, which involves the periodic exchange of elite (high-quality) solutions through the master node [143].

The master node retains a pool of elite solutions, each one detected by a different algorithm. The pool is asynchronously updated whenever a new elite solution is detected by one of the running algorithms. The new solution replaces the previous best solution found by the same algorithm. Then, the overall best solution is periodically forwarded to each algorithm. This solution (or perturbations of it) can be used either to initiate a new trajectory in trajectory-based methods, or it can be embedded in the population of a population-based method, possibly replacing its worst member. This mechanism essentially promotes the alleviation of search stagnation.

An additional intensification strategy that can be useful is path-relinking, which is used to explore paths connecting previously detected elite solutions [130]. In our problem of interest, a path-relinking procedure is induced according to a user-defined probability. During path-relinking, the best solution between an algorithm's own elite solution (initial point) and an acquired one (target point) is discovered. More specifically, the permutation of the initial point that has the lowest Hamming distance from the target point is detected and either triggers a new trajectory or becomes a new member of the population, depending on the type of the algorithm.

6.1.5 First Algorithm Portfolio Approach

The algorithm portfolio model with trading-based resource allocation was successfully applied on the circulant weighing matrices problem in [143]. The specific model treats the detected solutions as stocks that can be bought and sold by the constituent algorithms, using part of their allocated running time as currency. The full operational details of this model are exposed in Sect. 4.2.1. The specific

Table 6.1 Parameter setting for the considered algorithms [143]

Algorithm	Parameters
Tabu search (TS)	Tabu list size: $l = 48$
Variable neighborhood search (VNS)	Number of neighborhoods: $k_{\max} = 2$
Differential evolution (DE)	Population size: $N = 100$
	Mutation and crossover parameters: $F = 0.7, CR = 0.3$
Particle swarm optimization (PSO)	Swarm size: $N = 100$
	Constriction coefficient: $\chi = 0.729$
	Cognitive and social parameter: $c_1 = c_2 = 1.49$
	Neighborhood topology: ring of radius 1
Algorithm portfolio (AP)	Constants: $\lambda = 0.3, \theta = 0.05$
	Non-improvement cycles: $T_{\text{noimp}} = 5000$
Common parameters	Migration period: $T_{\text{mig}} = 100$ iterations

algorithm portfolio in [143] comprised different variants of tabu search, motivated by its promising performance in previous works [35].

In addition to the main algorithm portfolio model, a scheme that combines solutions was used in [143]. More specifically, the elite solution detected by a constituent algorithm was combined with its purchased solution through crossover and mutation operators. The two operators were inspired by genetic algorithms and work as follows: crossover is initially applied to keep components that are identical in the two solutions. Mutation is subsequently used to randomly perturb the components where the two solutions differ. The resulting point is used as a restart point for tabu search.

The targeted problem in [143] was the detection of CW(48, 36) matrices. The proposed algorithm portfolio was applied on the specific problem and compared against parallel implementations of the metaheuristics discussed in Sect. 6.1.4. The parameter configuration for each approach is reported in Table 6.1. All experiments were conducted on the `glacier` cluster of the Sharcnet[1] consortium using either 8 or 16 nodes, henceforth denoted as CPUs. In both cases, one of the nodes served as the master node, while the rest were slave nodes devoted to a single algorithm instance each. The total time budget in the experiments was either 12 or 24 hours, and it was equally distributed among the slave nodes. The resulting combinations of nodes-budget are henceforth denoted as 8/12, 8/24, 16/12, and 16/24, respectively.

In order to statistically validate the performance of the algorithms, 25 independent experiments were conducted per algorithm-nodes-budget combination, recording the best solution detected within the available running time. The considered performance criteria were the number of successful experiments, i.e., the experiments where an optimal solutions was detected, as well as the time required for the detection of the optimal solution. Moreover, another interesting criterion was the number of unique solutions detected by each algorithm in the 25 experiments. Note that a specific circulant weighing matrix can be defined by its first row but

---

[1] https://www.sharcnet.ca

Table 6.2 For each combination of algorithm, time budget, and number of CPUs, the table reports the corresponding number and percentage of successful experiments (suc) over a total of 25 experiments, along with the number and percentage of unique solutions (uni) detected with respect to the total number of solutions found by the algorithm [143]

Algorithm	Budget		CPUs	
			8	16
TS	12	suc	14 (56.0%)	22 (88.0%)
		uni	7 (50.0%)	5 (22.7%)
	24	suc	20 (80.0%)	24 (96.0%)
		uni	4 (18.2%)	6 (25.0%)
VNS	12	suc	3 (12.0%)	4 (16.0%)
		uni	1 (33.3%)	1 (25.0%)
	24	suc	5 (20.0%)	6 (24.0%)
		uni	2 (40.0%)	2 (33.3%)
DE	12	suc	8 (32.0%)	10 (40.0%)
		uni	1 (12.5%)	2 (20.0%)
	24	suc	15 (60.0%)	18 (72.0%)
		uni	3 (20.0%)	4 (22.2%)
PSO	12	suc	7 (28.0%)	11 (44.0%)
		uni	2 (28.6%)	3 (27.2%)
	24	suc	13 (52.0%)	17 (68.0%)
		uni	4 (30.8%)	5 (29.4%)
AP	12	suc	15 (60.0%)	25 (100.0%)
		uni	7 (46.7%)	10 (40.0%)
	24	suc	15 (60.0%)	25 (100.0%)
		uni	5 (33.3%)	9 (36.0%)

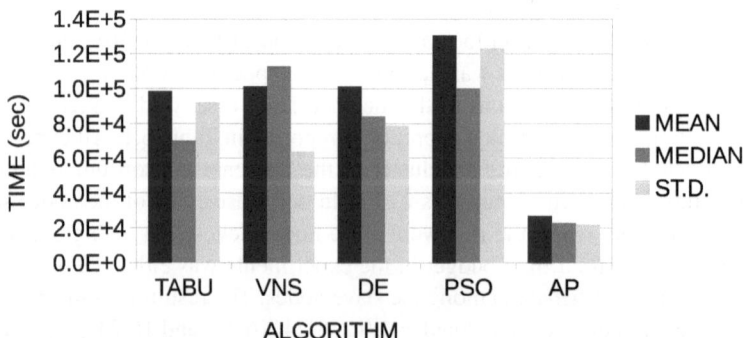

Fig. 6.1 Required running time (in seconds) per algorithm for the successful experiments over all nodes-budget combinations

also with all its right cyclic shifts. Thus, an algorithm can detect different global minimizers at each experiment, but eventually, they may all correspond to the same matrix. Obviously, algorithms that tend to detect different matrices in independent experiments are preferable.

A number of interesting results were reported in [143] and reproduced here in Table 6.2 and Fig. 6.1. Table 6.2 reports the number and percentage of successful

experiments and unique solutions for each algorithm and nodes-budget combination [143]. Considering the distinct metaheuristics, tabu search evidently
outperformed the rest of the algorithms both in successes and unique solutions.
Obviously, its exhaustive neighborhood search along with the inherent hill-climbing
capabilities of the algorithm and the employed cooperative parallel scheme have
resulted in a superior implementation for the studied problem.

Table 6.2 reveals also that, despite its overall effectiveness, tabu search performed
worse than the trading-based algorithm portfolio in almost all experimental configurations. The 8/24 case stands as exception, where tabu search was superior in
terms of successes. However, even in this case, the algorithm portfolio detected
5 unique solutions in 15 successful experiments against 4 unique solutions in 20
successful experiments of tabu search. Moreover, the algorithm portfolio was the
only approach that achieved 100% successes in any experimental configuration
(cases 16/12 and 16/24). All these evidence experimentally verify the benefits
gained from the special trading scheme incorporated into the algorithm portfolio
against the simple cooperative scheme adopted by the parallel implementations of
the individual metaheuristics.

Table 6.2 reveals also another performance aspect of the algorithms, namely,
the performance boost achieved when the number of CPUs or the running time
is doubled. Furthermore, it reveals that doubling the running time for the individual metaheuristics is more beneficial than doubling the number of CPUs in
most experimental cases. Surprisingly, this trend is not verified for the algorithm
portfolio. In that case, higher number of CPUs appears to have more significant
impact in the results than extending the running time. This behavior can be better
interpreted by the evidence displayed in Fig. 6.1, which illustrates the required
running time per algorithm only for the successful experiments over all experimental
configurations. The running time required by the algorithm portfolio approach to
detect an optimal solution is much lower than the corresponding time needed by
the individual metaheuristics. Therefore, prolonging the running of the algorithm
portfolio does not result in subsequent performance improvement. On the contrary,
providing additional slave nodes enhances its search capability, leading to increased
number of successes. As reported in [143], the observed differences of the running
times were statistically verified through Wilcoxon rank-sum tests [165]. The tests
showed that only algorithm portfolios achieved statistically significant differences
in running time with the rest of the algorithms.

Overall, the preliminary attempts of solving circulant weighing matrices problems with algorithm portfolios were very promising, cultivating the ground for
further developments. Given that the only difference between the tested algorithm
portfolio and the rest of the algorithms was its special solution trading scheme
and the subsequent dynamic allocation of the computation budget, we can infer
that the employed parallel algorithm portfolio framework can achieve significant
performance gains. The reader is referred to [143] for further details on the specific
application.

6.1.6 Second Algorithm Portfolio Approach

The second approach of the circulant weighing matrices problem with algorithm portfolios appeared in [140]. It was based on the algorithm portfolio model with performance forecasting, which is thoroughly described in Sect. 4.2.2. The specific algorithm portfolio was enriched with diverse algorithms and a different resource allocation mechanism than the one in the previous section.

More specifically, the metaheuristics that composed the algorithm portfolio included the promising tabu search and variable neighborhood search methods, as well as the iterated local search method (all presented in Chap. 1). For the employed metaheuristics, four state-of-the-art neighborhood search operators for permutation optimization were considered [176]. Specifically, let $\mathbf{x} = (x_1, x_2, \ldots, x_n)$ be a sequence that defines a $CW(n, k^2)$ matrix (i.e., $\mathbf{x}$ is its first row), and let x_i, x_j, $i < j$ be two components of $\mathbf{x}$. Then, the following local search operators can be used:

(a) Interchange (O_1): swaps x_i and x_j.
(b) Relocate (O_2): relocates x_i immediately after x_j.
(c) Or-opt (O_3): relocates x_i and x_{i+1} between x_j and x_{j+1}.
(d) 2-opt (O_4): interchanges x_{i+1} and x_j and reverses the order of all components between them.

The neighborhood of sequence $\mathbf{x}$ includes all its permutations produced by one application of the corresponding operator on $\mathbf{x}$. Note that operators O_2, O_3, and O_4 shift or reverse entire subsequences of $\mathbf{x}$, with O_4 having the strongest expected effect.

Another important issue is the selection between breadth or depth in probing the neighborhood, i.e., whether the algorithm will exhaustively probe each neighborhood or, alternatively, select the first point of improvement. In the first case, also known as neighborhood-best, deepest descent is performed ensuring the detection of a local minimizer if one exists in the neighborhood. The second case, also known as first-best, moves rapidly between neighborhoods thereby sparing a significant amount of function evaluations. However, this comes at the cost of probably overshooting the minimizer. These two approaches are frequently applied together in relevant applications, while their relative performance is problem-dependent.

The algorithm portfolio proposed in [140] comprised seven constituent algorithms, namely, two variants of tabu search, two variants of variable neighborhood search, and three variants of iterated local search. The two tabu search variants adopted the O_1 neighborhood operator and tabu list size $l = n/2$. Their difference lied in the strategy for exploring a neighborhood. In particular, while the one variant utilized the neighborhood-best approach, the other one used the first-best approach.

The same hold for the variable neighborhood search variants, where one variant was based on neighborhood-best search and the other one on first-best search strategy. Also, both variants utilized all four local search operators presented above. Iterated local search explores the search space by applying a local search procedure

Table 6.3 Experimental configuration in [140]

Description	Variant	Properties/values
Test problems		CW(33, 25), CW(42, 16), CW(48, 36), CW(52, 36)
		CW(57, 49), CW(62, 16), CW(70, 16), CW(78, 9)
		CW(84, 16), CW(96, 9), CW(112, 16), CW(130, 9)
Tabu search (TS)	A_1	Tabu list size: $l = n/2$ + neighborhood-best
	A_2	Tabu list size: $l = n/2$ + first-best
Variable neighborhood search (VNS)	A_3	Neighborhood-best
	A_4	First-best
Iterated local search (ILS)	A_5	Neighborhood operator: O_1 + neighborhood-best
	A_6	Neighborhood operator: O_2 + neighborhood-best
	A_7	Neighborhood operator: O_3 + neighborhood-best
Algorithm portfolio (AP)	NoF	Standard model
	SMA	Simple moving average (1-step lag)
	SES	Simple exponential smoothing, $\alpha = 0.3$
	LES	Linear exponential smoothing, $\alpha = 0.3$, $\beta = 0.8$
Common AP parameters		Number of batches: 100
		Number of CPUs: 36 (1 master and 35 slave nodes)
		Function evaluations: 10^{10}
		Experiments per case: 100

that typically considers a single neighborhood operator. Therefore, three variants of this algorithm were considered, using the O_1, O_2, and O_3 operator, respectively. Also, the neighborhood-best search was adopted in all cases.

The seven algorithm variants were selected to construct four heterogeneous algorithm portfolios in [140]. Three of the portfolios were based on the performance forecasting model presented in Sect. 4.2.2. Each one was based on a different forecasting technique, namely, simple moving average [69, 96], simple exponential smoothing [94, 96, 134], and linear exponential smoothing [96]. Another portfolio adopted the standard model with equal distribution of the computation resources among the constituent algorithms and without information exchange among them. This portfolio was used as the baseline approach in comparisons.

Twelve circulant weighing matrix problems of sequence length ranging from 33 up to 130 and weights ranging 9 up to 49 were used for benchmarking in [140]. All selected problems are reported as solvable in the relevant literature [13], and they constitute hard permutation optimization tasks as they have huge search spaces. The complete configuration framework of the experiments is reported in Table 6.3.

The experiments were conducted on a heterogeneous Beowulf cluster[2]using 36 CPUs (1 master node and 35 slave nodes). The master node accomplished all bookkeeping procedures, while the rest of the nodes were devoted to the algorithms. Initially, the 35 CPUs were equally distributed among the seven algorithms of

---

[2]The cluster was composed of 6th and 7th generation Intel® Core i7 processors.

the algorithm portfolio, each one occupying 5 CPUs. For the NoF portfolio, this allocation remained unchanged throughout each experiment. On the contrary, the rest of the portfolios dynamically modified the CPUs-algorithm assignment based on the forecasted performance, as analyzed in Sect. 4.2.2.

Each algorithm portfolio was tested in 100 independent experiments on each test problem. The available computation budget per experiment was 10^{10} function evaluations. Moreover, each constituent algorithm was initiated from a new random position if no improvement of its best solution occurred after 10^3 function evaluations. Finally, the allocation of the function evaluations to the algorithms was made in 100 batches.

At each experiment, the number of function evaluations required by the algorithm portfolio to detect an optimal solution was recorded. The distribution of the recorded values over 100 experiments is graphically illustrated in the boxplots of Fig. 6.2. Careful inspection of the boxplots reveals that the performance of the algorithm portfolios is heavily dependent on the corresponding problem. Moreover, the overlapping confidence intervals of the medians of each sample, as illustrated by the corresponding notches of the boxes, indicate that some of the observed differences may be statistically insignificant. For these reasons, the comparisons between the algorithm portfolios were based on Wilcoxon rank-sum tests [165] conducted for each pair of portfolios and test problem. Portfolio A was considered superior to another portfolio B on a specific test problem, if their recorded results had statistically significant difference and A achieved smaller average number of function evaluations. This case was counted as a win for A and loss for B. The absence of statistical significance was counted as a tie for both A and B. Given the number of the competing portfolios, three statistical comparisons were conducted for each portfolio and test problem, resulting in a total of 36 comparisons.

Table 6.4 summarizes the number of wins, losses, and ties per algorithm portfolio over all test problems. A first reading of the table shows that the performance of the three portfolios with forecasting was proportional to the amount of information incorporated in their forecasting technique. In particular, the algorithm portfolio with linear exponential smoothing outperformed the rest of the approaches, scoring 25 wins over 36 comparisons, while it was followed by the simple exponential smoothing approach with 20 wins. These two portfolios dominated the simplistic simple moving average approach as well as the standard algorithm portfolio without forecasting. Between these two portfolios, the simple moving average approach, which simply follows the time series with 1-step delay, was marginally better than the standard algorithm portfolio in terms of wins, exhibiting also a 24% reduced number of losses [140].

Additionally to this evidence, a special ranking scheme of the competing portfolios was proposed in [140]. More specifically, for each test problem, the four competing portfolios were ranked according to their number of wins based on the statistical comparisons among them for the specific problem. Thus, better portfolios were ranked first, followed by inferior ones. A first tiebraker criterion was the number of losses, in order to promote portfolios with smaller numbers of losses. A second tiebraker criterion was the average number of function evaluations required

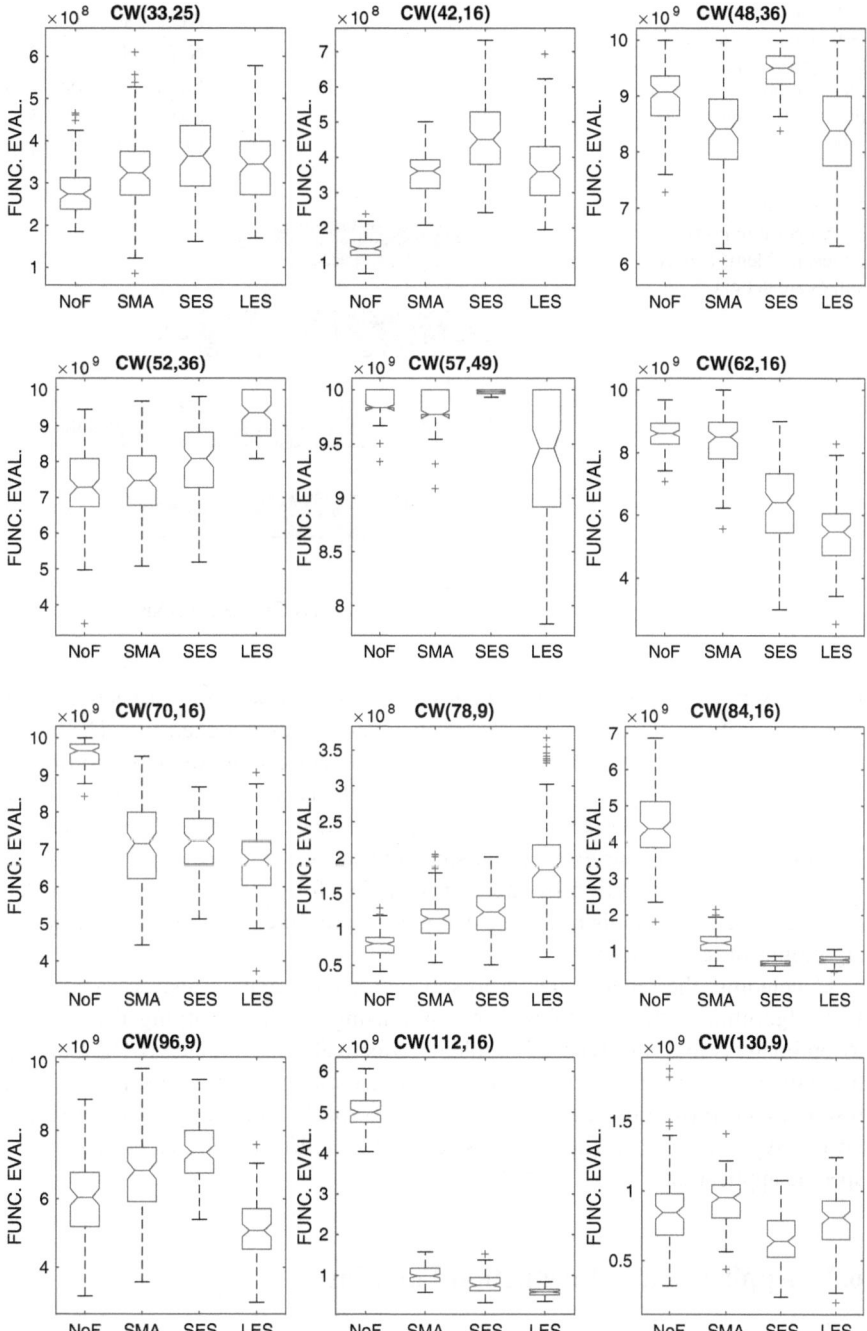

Fig. 6.2 Boxplots of the required number of function evaluations per problem and algorithm portfolio in 100 experiments

Table 6.4 Number of wins
(+), losses (−), and ties (≈)
per algorithm portfolio over
all test problems derived by
the statistical significance
tests [140]

AP	+	−	≈
NoF	9	25	2
SMA	10	19	7
SES	20	13	3
LES	25	7	4

Fig. 6.3 Average rank of
each algorithm portfolio over
all test problems (lower
values are better)

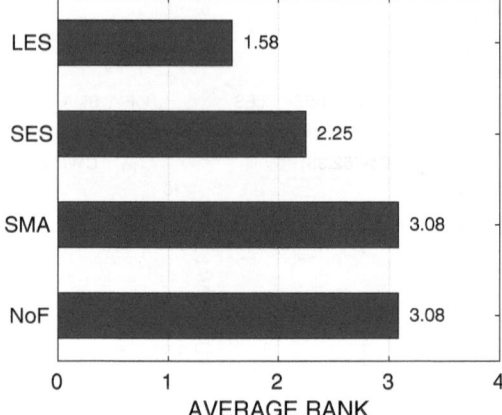

for the specific problem, which promotes more efficient portfolios. Eventually, for
each test problem, a unique rank between 1 and 4 was assigned to each one of
the algorithm portfolios. Obviously, smaller ranks correspond to better portfolios.
These ranks were then averaged over all test problems, and the obtained results are
illustrated in Fig. 6.3. Clearly, the evidence is aligned with the previous findings.
The superiority of linear exponential smoothing was attributed to the use of both
level and trend information by the corresponding forecasting mechanism, which
managed to efficiently distribute the computation resources among the constituent
algorithms of the portfolio.

Concluding, the second attempt to solve circulant weighing matrices problems
with algorithm portfolios offered very promising results, verifying their ability
to optimally solve this type of problems effectively and efficiently. Monitoring
the online performance of the portfolio and accordingly allocating the available
resources has proved to be a very promising approach, motivating further research
on this type of solvers. The reader is referred to [140] for further details on the
specific application.

6.2   Application in Production Planning

Manufacturing companies are considered top waste producers. Current environmental policies worldwide promote the reduction of material waste through various
procedures. One of them is the recovery of already used products, whenever

possible. This initiative has been motivated by the necessity for reduction of the environmental impact [120], potential financial benefits [175], as well as the economic growth through the creation of new job opportunities [168].

One approach to reduce waste material is the transformation of already used products into like-new ones, a process widely known as remanufacturing [108]. More specifically, remanufacturing involves the following procedures:

(a) Dismantling the product and replacing or repairing its problematic components.
(b) Testing the modified parts and the whole product to ensure that they meet design specifications.

Remanufactured products are guaranteed to have equivalent quality to new ones, but they are usually sold at competitive prices in order to attract buyer's interest.

Economic lot sizing refers to production planning over a discrete and finite planning horizon, where demand is assumed to be dynamic and deterministic. The solution of problems of this type has been an active research area for many decades [31, 32, 161]. Recently, economic lot sizing with remanufacturing options has received increasing attention as an alternative to the standard manufacturing process [57, 121, 122, 129]. This problem considers a single-type production system that meets a dynamic and deterministic demand over a finite, discrete planning horizon by manufacturing new products and remanufacturing returned ones.

In order to meet the demand at each time period, the manufacturer shall determine the number of new items that need to be produced as well as the fraction of the returned items that need to be remanufactured. Manufactured and remanufactured items are stored in separate inventories. Additionally, fixed costs are incurred for manufacturing or remanufacturing a product, and holding costs are charged for keeping new and returned items in the inventory. The underlying optimization problems are of mixed integer linear programming type, offering a challenging approach for metaheuristics as well as algorithm portfolios.

6.2.1   Problem Formulation

The economic lot sizing problem with remanufacturing assumes a finite planning horizon composed of T time periods. In our scenario, a manufacturer sells a single-type product to satisfy a dynamic and deterministic demand D_t at each time period, $t = 1, 2, \ldots, T$. Each product is obtained either from a manufacturing process, in which new items are created, or from a remanufacturing process in which a fraction R_t of returned items are recovered and sold as new.

Until their sale, recovered products are retained in an inventory of recoverable items, assuming a holding cost h^R per unit time. At each time period t, a number of z_t^R and z_t^M products are remanufactured and manufactured, respectively, and then forwarded to a serviceables inventory with a holding cost h^M per unit time. Naturally, setup costs are incurred by the remanufacturing and manufacturing process, denoted as K^R and K^M, respectively.

The main optimization objective in such production systems is to determine the exact number of remanufactured and manufactured items per time period, in order to minimize the sum of the incurring setup and holding costs under a number of operational constraints. The considered problem can be modeled as a mixed integer linear programming problem, whose cost function is defined as follows [136]:

$$C = \sum_{t=1}^{T} \left(K^R \gamma_t^R + K^M \gamma_t^M + h^R y_t^R + h^M y_t^M \right),$$ (6.6)

where:

$$\gamma_t^R = \begin{cases} 1, & \text{if } z_t^R > 0, \\ 0, & \text{otherwise,} \end{cases} \qquad \gamma_t^M = \begin{cases} 1, & \text{if } z_t^M > 0, \\ 0, & \text{otherwise,} \end{cases}$$ (6.7)

for $t = 1, 2, \ldots, T$ are binary variables denoting the initiation of a remanufacturing or manufacturing lot, respectively. The inventory levels of items that can be remanufactured or manufactured in period t are denoted by y_t^R and y_t^M, respectively. A number of operational constraints accompany the model [136]

$$(1) \quad y_t^R = y_{t-1}^R + R_t - z_t^R,$$ (6.8)

$$(2) \quad y_t^M = y_{t-1}^M + z_t^R + z_t^M - D_t,$$ (6.9)

$$(3) \quad z_t^R \leq Q \, \gamma_t^R,$$ (6.10)

$$(4) \quad z_t^M \leq Q \, \gamma_t^M,$$ (6.11)

$$(5) \quad y_0^R = y_0^M = 0,$$ (6.12)

$$(6) \quad \gamma_t^R, \gamma_t^M \in \{0, 1\},$$ (6.13)

$$(7) \quad y_t^R, y_t^M, z_t^R, z_t^M \geq 0,$$ (6.14)

$$t = 1, 2, \ldots, T.$$

Constraints (1) and (2) guarantee balanced inventories for both returned and serviceable products, while constraints (3) and (4) assure that fixed costs are paid whenever a new lot is initiated. The constant Q is a sufficiently large positive number, whose value is suggested to be equal to the total demand of the planning horizon [136]. Finally, constraints (5)–(7) assert that inventories are initially empty, determine the domain of the indicators, and guarantee that the integer variables receive non-negative values. The decision variables of the optimization problem are the number of remanufactured items, z_t^R, and manufactured items, z_t^M, for each time period $t = 1, 2, \ldots, T$. Thus, considering a planning horizon of T periods, the dimension of the corresponding optimization problem is $n = 2T$. For additional information on the studied problem, the reader is referred to [103, 136, 154].

6.2.2 Application of Algorithm Portfolios

Metaheuristics have been habitually used in production systems optimization [70, 110, 148]. The economic lot sizing problem with remanufacturing has been tackled by population-based methods [103, 123, 124] as well as local search metaheuristics [138]. The emerging production problems are usually of integer or mixed integer nature; hence, combinatorial local search algorithms are directly applicable. On the other hand, several population-based methods that are primarily designed to address continuous problems require modifications in problem formulation or in their operators to become applicable. The reader is referred to Sect. 6.1.4 for relevant discussion.

In [142], the potential of applying the algorithm portfolio with trading-based resource allocation on the studied problem was investigated. The portfolio consisted of two local search methods, namely, tabu search and iterated local search, as well as two population-base metaheuristics, namely, particle swarm optimization and differential evolution. Benchmarking was based on the test suite developed by T. Schulz [136] and includes a full factorial study of various problem instances that share a planning horizon of $T = 12$ periods as well as the parameters reported in Table 6.5. The test suite contains 6480 test problems, and it was made available after personal communication of the co-author of [103] I. Konstantaras with its developer.

Table 6.6 reports the parameter configuration of the algorithm portfolio and its constituent algorithms in [142]. The primary experimental objective was to investigate the portfolio's performance in terms of solution quality and compare it to its constituent algorithms as well as against the best-performing variant SM_4^+ of the state-of-the-art Silver-Meal heuristic tested in [136]. A predefined time budget T_{tot} was allocated to each constituent algorithm, and performance was quantified in terms of the gap percentage of the attained solution value to the known global minimum, which was computed using the CPLEX11 software and reported in [136].

Table 6.5 Problem parameters of the considered test suite [136]

Parameter	Value(s)
Dimension	$n = 24$
Setup costs	$K^M, K^R \in \{200, 500, 2000\}$
Holding costs	$h^M = 1, h^R \in \{0.2, 0.5, 0.8\}$
Demand for period t	$D_t \sim N(\mu_D, \sigma_D^2)$
	$\mu_D = 100$
	$\sigma_D^2 = 10\%$ of μ_D (small variance)
	$\sigma_D^2 = 20\%$ of μ_D (large variance)
Returns for period t	$R_t \sim N(\mu_R, \sigma_R^2)$
	$\mu_R \in \{30, 50, 70\}$
	$\sigma_R^2 = 10\%$ of μ_R (small variance)
	$\sigma_R^2 = 20\%$ of μ_R (large variance)

Table 6.6 Parameters of the studied algorithms [142]

Algorithm	Parameters
Particle swarm optimization (PSO)	Model: lbest with ring topology
	Swarm size: 60
	Constriction coefficient: $\chi = 0.729$
	Cognitive and social parameter: $c_1 = c_2 = 2.05$
Differential evolution (DE)	Population size: 60
	Operator: DE2
	Mutation parameter: $F = 0.7$
	Crossover parameter: $CR = 0.3$
Tabu search (TS)	Tabu list size: $l = 24$
Algorithm portfolio (AP)	Number of constituent algorithms: 4
	Total running time: 300 seconds
	Constants: $\lambda = 0.1$, $\theta = 0.05$

Table 6.7 Gap percentage over all test problems [142]

Algorithm	Mean	St.D.	Max
SM_4^+	2.2	2.9	**24.3**
PSO	4.3	4.5	49.8
DE	3.3	5.1	31.9
TS	51.6	33.4	255.5
ILS	80.3	54.3	450.8
AP	**1.9**	**2.8**	35.6

Table 6.7 reports the average, standard deviation, and maximum value of the achieved gap percentage per algorithm over all the 6480 test problems. Table 6.8 reports the same information per class of test problems. In both cases, best values per case are boldfaced. The experimental evidence shows that, overall, the algorithm portfolio outperforms the rest of the algorithms, achieving the lowest mean gap percentage 1.9%. The SM_4^+ heuristic is the second best approach with a mean gap percentage of 2.2%, followed by differential evolution with 3.3% and particle swarm optimization with 4.3%. A close inspection of Table 6.8 reveals also that the algorithm portfolio achieved the lowest gap percentage in 13 out of 16 problem classes. Interestingly, we can also see that the population-based algorithms were superior to the local search metaheuristics. This can be attributed to the fact that the considered problem requires rather an exploration-oriented approach to be solved effectively [142].

Moreover, it was observed that the trading among the constituent algorithms of the portfolio can further enhance solution quality. Deeper analysis in [142] revealed that the population-based algorithms were mainly solution sellers, in contrast to local-based algorithms that were mostly solution buyers. This evidence suggests that exploration-oriented approaches could detect better solutions than exploitation-oriented ones. Careful runtime analysis of the local search methods showed that they could hardly escape from local optima. Consequently, they were trying to mitigate

Table 6.8 Gap percentage per class of test problems [142]

Algorithm	Case	Mean	St.D.	Max	Case	Mean	St.D.	Max
SM_4^+	$\sigma_D^2=10\%$	2.1	2.8	**18.9**	$h^R=0.2$	1.7	2.5	**21.1**
PSO		4.4	4.6	49.8		4.5	5.2	49.8
DE		3.4	4.8	31.7		3.0	5.3	30.9
TS		50.9	33.2	200.2		45.0	26.4	255.5
ILS		79.7	54.2	450.8		94.8	67.2	450.8
AP		**1.8**	**2.6**	26.8		**1.5**	**2.5**	35.6
SM_4^+	$\sigma_D^2 = 20\%$	2.4	3.0	**24.3**	$h^R = 0.5$	2.3	3.0	**24.3**
PSO		4.1	4.5	48.3		4.3	4.5	45.5
DE		3.3	5.2	31.9		3.3	5.0	31.9
TS		52.4	33.5	255.5		50.8	32.1	202.1
ILS		80.9	54.4	421.5		77.6	48.2	261.1
AP		**2.0**	**2.9**	35.6		**1.9**	**2.8**	27.4
SM_4^+	$\mu_R = 30$	**1.2**	**1.8**	**12.1**	$h^R = 0.8$	2.8	3.0	**20.6**
PSO		3.5	3.1	45.5		4.0	3.9	42.9
DE		3.3	5.0	28.2		3.7	4.5	31.4
TS		37.2	23.9	255.5		59.1	39.0	235.2
ILS		70.6	46.5	336.8		68.4	40.8	211.4
AP		1.6	2.5	25.7		**2.2**	**3.0**	21.6
SM_4^+	$\mu_R = 50$	2.3	2.7	**16.2**	$K^M = 200$	**2.3**	**2.6**	**13.5**
PSO		4.1	4.0	34.0		4.0	3.1	45.5
DE		3.5	5.2	31.9		3.2	3.9	24.0
TS		50.8	27.3	153.6		39.1	27.3	255.5
ILS		83.7	53.0	364.2		62.6	64.0	450.8
AP		**2.0**	**2.9**	27.4		2.4	3.0	21.6
SM_4^+	$\mu_R = 70$	3.3	3.5	**24.3**	$K^M = 500$	2.1	2.5	12.8
PSO		5.1	5.9	49.8		4.5	4.1	27.5
DE		3.3	4.6	31.7		2.5	2.6	15.2
TS		66.9	39.8	235.2		67.9	33.1	197.1
ILS		86.5	61.1	450.8		62.0	40.6	278.1
AP		**2.0**	**2.9**	35.6		**1.8**	**2.4**	17.6
SM_4^+	$K^R = 200$	**1.9**	**2.1**	**11.8**	$K^M = 2000$	2.3	3.4	**24.3**
PSO		5.7	5.5	49.8		4.4	5.9	49.8
DE		3.8	4.0	24.0		4.3	7.1	31.9
TS		75.2	38.0	203.3		47.9	32.7	235.2
ILS		63.0	45.4	260.4		116.3	34.2	260.4
AP		3.0	3.3	21.6		**1.4**	**2.8**	35.6
SM_4^+	$K^R = 500$	3.4	3.2	19.1	$\sigma_R^2 = 10\%$	2.2	2.9	**21.1**
PSO		3.8	4.1	37.4		4.3	4.6	46.7
DE		1.8	2.0	11.2		3.4	5.0	31.4
TS		50.8	23.8	235.2		52.1	34.3	233.8

(continued)

Table 6.8 (continued)

Algorithm	Case	Mean	St.D.	Max	Case	Mean	St.D.	Max
ILS		62.4	37.8	244.8		80.4	54.4	450.8
AP		**1.3**	**1.7**	**11.6**		**1.8**	**2.7**	35.6
SM_4^+	$K^R = 2000$	1.4	2.9	**24.3**	$\sigma_R^2 = 20\%$	2.3	2.9	**24.3**
PSO		3.3	3.5	45.5		4.2	4.5	49.8
DE		4.4	7.1	31.9		3.3	4.9	31.9
TS		29.0	16.4	255.5		51.2	32.5	255.5
ILS		115.5	59.2	450.8		80.1	54.2	399.1
AP		**1.3**	**2.8**	35.6		**2.0**	**2.9**	25.6

this effect by buying solutions from the population-based algorithms. Considering this distinct buyer/seller role, it is concluded that the portfolio's constituent algorithms exhibited complementarity, which is a highly desirable property in algorithm portfolios [118, 160].

Summarizing, the application of algorithm portfolios proposed in [142] showed that the interplay of the constituent algorithms in the trading-based model can offer significant performance gains. Strong indications of complementarity between the algorithms were observed, verifying this important property in the specific model. The reader is referred to [142] for a thorough analysis of this application.

6.3 Application in Humanitarian Logistics

The term humanitarian logistics is used for the branch of logistics that specializes in coordinating the delivery of humanitarian aid to people affected by natural or man-made disasters. Quoting from [155], humanitarian logistics is responsible for:

> ...planning, implementing and controlling the efficient, cost-effective flow and storage of goods and materials, as well as related information, from point of origin to point of consumption for the purpose of alleviating the suffering of vulnerable people.

In general, humanitarian logistics has beneficial impact in various disaster relief situations [132, 159], by offering strategies that mitigate the negative effects of the disaster in terms of life loss and economic consequences. Additionally, it stores and processes a multitude of logistics data of past disaster incidents, promoting prompt reaction in future events. Therefore, it is generally perceived also as a crucial post-disaster analysis tool [67].

The increasing number of natural and human-caused disasters has resulted on a multitude of research works on humanitarian logistics over the past two decades [27, 83, 107, 111, 153]. Several studies have focused on modeling a variety of such problems [4, 157] that emerge in various planning stages of the disaster lifecycle, including preparedness, post-disaster, and recovery phases [112].

Moreover, diverse solver types, including both exact [1, 127, 133] and metaheuristic approaches [47, 60, 76, 102], have been proposed to solve the underlying optimization challenges.

Despite the social impact of humanitarian logistics, relevant literature is limited when compared to commercial logistics where the primary objective is the reduction of inherent costs [159]. Humanitarian logistics involve also different objectives than costs, thereby introducing new characteristics in the existing problems. For this reason, there is sheer need for sophisticated optimization models that incorporate essential aspects of these challenges, as well as optimization algorithms tailored to the requirements of the emerging problems.

6.3.1 Problem Formulation

In [76], a humanitarian logistics model with finite planning horizon of T time periods was proposed. The studied scenario included a set J of affected areas and a set I of dispatch centers. Relief products (commodities of different types) are transported via different modes of transportation and vehicle types from dispatch centers to affected areas under various traffic restrictions. The model assumes ground and aerial modes of transportation, as well as big and small types of vehicles. Henceforth, the set of commodities is denoted as C, the set of transportation modes is denoted as M, and the set of vehicles of mode $m \in M$ is denoted as O_m. Table 6.9 presents in detail all the model and decision variables used to formulate the studied problem along with their description.

The first optimization objective of the model is to specify the exact quantities s^t_{cijm} per commodity $c \in C$ that are delivered from dispatch center i to affected area j using vehicles of transportation mode m for each time period $t = 1, 2, \ldots, T$. A second objective is to determine the exact number v^t_{cijmo} of vehicles of type o and mode m that are used to transport the commodities for each time period t. The delivered quantities and the number of vehicles assume integer values.

The objective function of the minimization problem is defined as follows [76]:

$$\min \sum_{t \in T} \sum_{j \in J} \sum_{c \in C} \left[b_{cj} \left(d^t_{cj} - \sum_{i \in I} \sum_{m \in M} s^t_{cijm} - L^{t-1}_{cj} \right)^2 \right], \tag{6.15}$$

where b_{cj} denotes a constant that quantifies the importance of commodity c at affected area j. The model comes along with a number of operational constraints:

(1) $\quad L^0_{cj} = Y_{cj}, \quad \forall c \in C, \forall j \in J$ \hfill (6.16)

(2) $\quad L^t_{cj} = \sum_{i \in I} \sum_{m \in M} s^t_{cijm} - d^t_{cj} + L^{t-1}_{cj}, \quad \forall t \in T, \forall c \in C, \forall j \in J$ \hfill (6.17)

Table 6.9 Notation used in the proposed humanitarian logistics model [76]

Model variable	Description
T	Planning horizon
I	Set of dispatch centers (DCs)
J	Set of affected areas (AAs)
C	Set of commodities
M	Set of transportation modes
m	Index denoting the transportation mode (ground, air)
O_m	Set of vehicle types of transportation mode m
o	Index denoting the vehicle type (big vehicle, small vehicle)
b_{cj}	Importance weight of commodity c in AA j
w_c	Unit weight of commodity c
$volume_c$	Unit volume of commodity c
cap_{mo}	Capacity of type o, mode m vehicle
vol_{mo}	Volume capacity of type o, mode m vehicle
d_{cj}^t	Demand for commodity c in AA j at time period t
k_{ijm}^t	Traffic restriction for mode m vehicles traveling from DC i to AA j at time t
v_{imo}^t	Number of type o, mode m vehicles at DC i at time t
L_{cj}^t	Inventory level of commodity c in AA j at time t
Decision variable	**Description**
s_{cijm}^t	Delivered quantity of commodity c from DC i to AA j through transportation mode m at time t
v_{cijmo}^t	Number of type o, mode m vehicles used at period t to transport commodity c from DC i to AA j

$$(3) \quad \sum_{c\in C}\sum_{j\in J} s_{cijm}^t w_c \leqslant \sum_{o\in O_m} v_{imo}^t cap_{mo}, \quad \forall t\in T, \forall i\in I, \forall m\in M \qquad (6.18)$$

$$(4) \quad \sum_{c\in C}\sum_{j\in J} s_{cijm}^t vol_c \leqslant \sum_{o\in O_m} v_{imo}^t vol_{mo}, \quad \forall t\in T, \forall i\in I, \forall m\in M$$

$$\hspace{12cm} (6.19)$$

$$(5) \quad s_{cijm}^t \leqslant \min\left\{ \frac{\sum_{o\in O_m} v_{cijmo}^t cap_{mo}}{w_c}, \frac{\sum_{o\in O_m} v_{cijmo}^t vol_{mo}}{volume_c} \right\} \qquad (6.20)$$

$$(6) \quad \sum_{c\in C}\sum_{o\in O_m} v_{cijmo}^t \leqslant k_{ijm}^t, \quad \forall t\in T, \forall i\in I, \forall m\in M, \forall j\in J \qquad (6.21)$$

$$(7) \quad \sum_{c\in C}\sum_{j\in J} v_{cijmo}^t \leqslant v_{imo}^t, \quad \forall t\in T, \forall i\in I, \forall m\in M, \forall o\in O_m \qquad (6.22)$$

$$(8) \quad s_{cijm}^t, v_{cijmo}^t, L_{cj}^t \in \mathbb{N}, \quad \forall t, c, i, j, m, o \qquad (6.23)$$

(9) $\quad v^t_{imo}, d^t_{cj}, vol_{mo}, cap_{mo}, volume_c, w_c \in \mathbb{N}^+, \quad \forall t, c, i, j, m, o \qquad (6.24)$

(10) $\quad \sum_{c \in C} b_{cj} = 1, \quad b_{cj} \in [0, 1], \forall j \in J \qquad\qquad\qquad (6.25)$

Constraint (1) specifies the initial inventory level of commodity c at dispatch center j. Inventory balance that considers the demand of the commodity c and the replenishment quantity is determined by constraint (2). Constraints (3) and (4) ensure that capacity and volume are not higher than predefined values, respectively. Constraint (5) guarantees that the delivered quantity s^t_{cijm} does not exceed the defined upper limit. This constitutes a useful approach for bounding decision variables. Constraint (6) imposes traffic flow restrictions that bound the flow of vehicles. Such restrictions are anticipated in natural disasters, e.g., due to road infrastructure that is partially or completely damaged. Constraint (7) asserts that the number of vehicles transporting commodities does not exceed the total number of available vehicles. Eventually, constraints (8)–(10) determine domains of the decision variables and problem parameters.

6.3.2 Application of Algorithm Portfolio

In the relevant literature, the algorithmic power of various metaheuristics has been exploited to solve humanitarian logistics challenges [47, 60, 102, 157, 162]. The model presented in the previous section was studied in [76] and solved using the algorithm portfolio approach with trading-based resource allocation. The algorithm portfolio consisted of particle swarm optimization algorithm and the standard version, as well as an enhanced version of the differential evolution metaheuristic, all presented in Chap. 1.

The decision variables of the studied problem are all integer. Thus, the employed population-based metaheuristics assumed rounding of their real-valued vectors to the nearest integer [76], an approach that has been successfully used in similar problems [116, 123]. More specifically, the algorithms operate in the real-valued search space to produce candidate solution vectors which, prior to their evaluation, are rounded to nearest integer vectors, component wisely. The rounded vectors are retained in the population of the differential evolution approaches, as well as in the best positions (not the swarm) of the particle swarm optimization algorithm [76].

An additional issue that needed to be addressed was constraint handling. This issue was tackled through the commonly used penalty function method [169], combined with the following set of preference rules between feasible and infeasible solutions:

(a) Between two infeasible candidate solutions, the one that violates the smallest number of constraints is selected.

(b) Between a feasible and an infeasible candidate solution, the feasible one is always preferred.
(c) Between two feasible candidate solutions, the one with the smallest objective value is considered.

These rules are frequently used with population-based metaheuristics in constrained optimization problems [116]. The employed penalty function P in [76] was defined as follows:

$$P(\mathbf{x}) = f(\mathbf{x}) + \sum_{i \in VS(\mathbf{x})} |V(i)|, \qquad (6.26)$$

where $f(\mathbf{x})$ is the actual objective function value of candidate solution $\mathbf{x}$, $V(i)$ encapsulates the relative violation degree of the i-th constraint, and $VS(\mathbf{x})$ is the set of constraints violated by $\mathbf{x}$. Note that if no constraint violation is introduced by $\mathbf{x}$, the penalty value $P(\mathbf{x})$ is equal to the original objective function value $f(\mathbf{x})$.

The experimental setting in [76] considered three commodity types, namely, water, medicines, and food. The first two were assumed to be more important than the third commodity type in the studied model, and hence, they were assigned higher importance scores. The relevant information for each commodity is summarized in Table 6.10. Both ground and aerial transportation vehicles, i.e., trucks and helicopters, were considered for the transportation of commodities from two dispatch centers, denoted as DC_1 and DC_2, to the affected areas. The vehicles were of two types, namely, small ones (denoted as type I) and large ones (denoted as type II). Table 6.11 reports the number of vehicles per transportation mode and dispatch center, as well as the corresponding capacity and volume information per vehicle type.

The data were based on real-world values (e.g., palettes of water bottles, typical transportation boxes for medication, etc.) as reported in [76], and ten problem instances, henceforth denoted as $P1$–$P10$, were generated for benchmarking. The

Table 6.10 Commodities and their properties in the experimental setting of [76]

	Water	Medicines	Food
Importance score	0.35	0.35	0.30
Unit weight (in kg)	650	20	200
Unit volume (in m^3)	1.44	0.125	0.60

Table 6.11 Number of vehicles, capacity, and volume information for vehicle type I (small) and type II (big)

	Transportation mode			
	Ground		Air	
	I	II	I	II
Number of vehicles at DC_1	4	5	1	1
Number of vehicles at DC_2	4	5	1	1
Load capacity (ton)	3	10	4	9
Load volume (m^3)	20	44	35	75

dimension of all test problems was $n = 144$. The generated instances were optimally solved using the commercial ILOG CPLEX® solver.

Extensive experiments were conducted using (i) each one of the employed metaheuristics, individually, (ii) algorithm portfolios consisting of pairs of metaheuristics, and (iii) an algorithm portfolio consisting of all three algorithms. The analysis included the best model of particle swarm optimization with ring topology of radius $r = 1$ and default parameters $\chi = 0.729$ and $c_1 = c_2 = 2.05$, as well as a multitude of differential evolution variants. More specifically, the five basic mutation operators presented in Sect. 1.3.2 were considered in combination with all parameter values $F \in [0, 2]$ and $CR \in [0, 1]$, with step size 0.05. Additionally, the enhanced operator defined in Sect. 1.3.3 with the same parameters combinations was used [76].

All algorithms assumed the same population size $N = 150$. This size was shown to be the most promising value for the considered problem dimension in a preliminary experimental phase dedicated to identify the most promising variant of each constituent algorithm. More specifically, the preliminary analysis revealed that the DE2 operator with parameters

$$F = F_1 = F_2 = 0.4, \quad CR = 0.05,$$

was the superior setting for the differential evolution variants [76].

In a second phase of experiments, the most promising variant of each algorithm was applied to the studied problem [76]. In particular, 30 independent experiments were conducted per problem instance to enable statistical comparisons among the considered solvers. Each algorithm was assigned a running time of 10 minutes in order to render its results comparable to that of CPLEX. For each independent experiment, the best solution $\mathbf{x}^*$ detected by the corresponding algorithm, along with its value $f^* = f(\mathbf{x}^*)$, was recorded. Eventually, the absolute solution error from the global minimum f_{cplex}^* detected by CPLEX was calculated as follows:

$$\epsilon^* = \left| f_{\text{cplex}}^* - f^* \right|.$$

The acquired error values were averaged over the 30 independent experiments in order to statistically analyze the performance of each algorithm. All acquired results are summarized in Table 6.12. Figure 6.4 illustrates the success rate of the most promising algorithms, i.e., the percentage of experiments where the algorithm succeeded to reach the optimal solution within the available running time.

A first inspection of the results offers interesting conclusions. As for the stand-alone approaches, Table 6.12 shows that the enhanced differential evolution outperformed both the standard differential evolution and the particle swarm optimization algorithm. This is a direct consequence of using its special mutation operator and a restart mechanism of mild perturbations (see Sect. 1.3.3), which enhance the diversification capabilities of the algorithm.

Table 6.12 Statistics of solution error values for all algorithms, averaged over all test problems. Best values per column are boldfaced. The constituent algorithms of the algorithm portfolios are given in parentheses [76]

Algorithm	Mean	St.D.	Min	Max
Particle swarm optimization (PSO)	513.80	235.85	197.00	2442.20
Differential evolution (DE)	63.31	40.45	26.97	160.21
Enhanced dofferential evolution (eDE)	3.54	3.42	0.29	11.80
Algorithm portfolio (PSO+DE)	52.28	31.11	27.01	129.42
Algorithm portfolio (PSO+eDE)	4.14	3.99	0.16	13.77
Algorithm portfolio (DE+DE)	59.65	55.36	21.15	193.81
Algorithm portfolio (DE+eDE)	0.76	**0.91**	**0.00**	2.91
Algorithm portfolio (eDE+eDE)	**0.75**	0.85	**0.00**	**2.27**
Algorithm portfolio (PSO+DE+eDE)	0.84	1.18	**0.00**	3.74

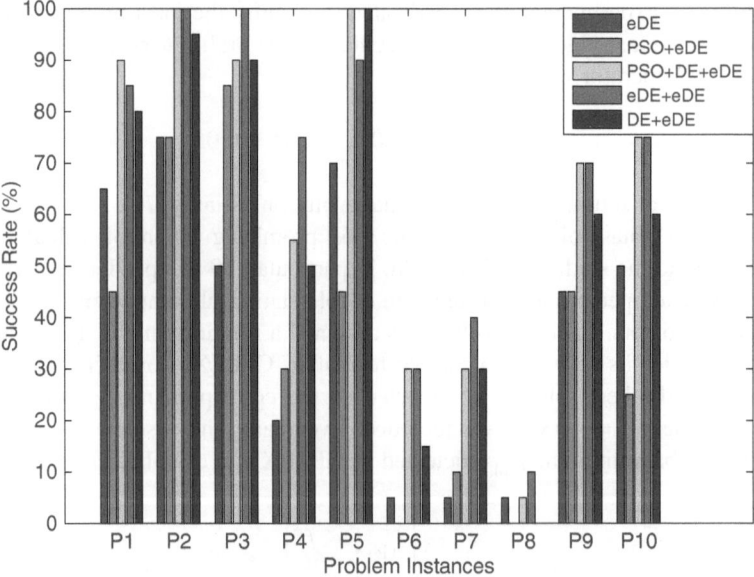

Fig. 6.4 Success rates of the most promising algorithms per problem. Symbol "+" denotes algorithm portfolios consisting of the corresponding approaches [76]

The superiority of the enhanced differential evolution is the main reason for the effectiveness of the algorithm portfolios that include this particular metaheuristic approach. More specifically, Table 6.12 suggests that the homogeneous algorithm portfolio composed of two instances of the enhanced differential evolution (denoted as eDE+eDE) as well as the heterogeneous algorithm portfolio composed of one standard and one enhanced instance of differential evolution (denoted as DE+eDE) outperformed all other approaches. These two approaches achieved the smallest average solution errors, equal to 0.75 and 0.76, respectively. However,

Fig. 6.4 reveals that the homogeneous eDE+eDE portfolio performed better than the DE+eDE for P6–P8, which were evidently the most difficult ones in terms of success rate. Moreover, the zero minimum ("Min") values in Table 6.12 show that algorithm portfolios based on enhanced differential evolution succeeded in optimally solving several problem instances.

An analysis of the solution-purchase frequency between the algorithms of the competing portfolios showed that, especially for these complex problems, the constituent algorithms of the portfolios exchanged large numbers of solutions in order to diversify their search because they experienced severe difficulties in reaching the optimal solution [76]. It shall be noted that the running time per algorithm in the case of the PSO+DE+eDE portfolio was lower than the rest because in the first case, the total time is distributed among three algorithms instead of two. Therefore, since particle swarm optimization was less effective than the enhanced differential evolution on the studied problems, the time assigned to the PSO+DE+eDE portfolio was obviously inadequate.

Summarizing, the application studied in [76] showed that the algorithm portfolios with trading-based model can be efficient solvers for humanitarian logistics problems such as the presented model. Also, it underlines that the number and selection of constituent algorithms play a crucial role in performance. The reader is referred to [76] for a more thorough presentation.

6.4 Synopsis

Effectiveness and efficiency of algorithm portfolios have been identified on several challenging problems spanning diverse application areas. In this chapter, we presented three such problems from the fields of combinatorics and operations research. For each problem, we provided a compact description and its mathematical formulation. Insight on the application of metaheuristics and algorithm portfolios with sophisticated resource allocation mechanisms was offered. Finally, the reported experimental results convince about the ability of the presented algorithm portfolio implementations to achieve remarkable performance compared to their constituent algorithms, as well as against other parallel approaches or specialized heuristics.

Chapter 7
Epilogue

Algorithm portfolios are continuously established as powerful algorithmic schemes in the relevant literature. From their introduction in the pioneering work of B.A. Huberman, R.M. Lukose, and T. Hogg [68] until today, they count an increasing number of publications and more than 1,300 relevant results in Google Scholar ®. This is indicative of their current dynamic and potential to occupy a position among the state-of-the-art in computational optimization.

The computational power of algorithm portfolios stems from their two essential components: their constituent algorithms and their resource allocation schemes. Both these components have attracted ongoing research interest for decades in fields such as artificial intelligence and operations research. Despite the large number of published works for each topic, individually, there is still fertile ground for new ideas and developments that will successfully address the challenges presented in the previous chapters.

In the present book, we tried to cover the basics of algorithm portfolios and review some of the most influential works according to our experience in the field. The covered topics included all critical issues in algorithm portfolios, namely:

(a) Selection of the constituent algorithms.
(b) Resource allocation schemes.
(c) Efficient implementation models.
(d) Application-related peculiarities.

Our interest focused on recent developments that appeared in the relevant literature in the past 10 years, especially those related to the metaheuristic optimization literature. Inevitably, in such vibrant research topics that dynamically evolve day by day, we may neglected some developments. We hope that these shortcomings are limited and nonessential for the introduction of the non-expert reader to the field.

Our experience in designing and applying algorithm portfolios on diverse applications has revealed the complexity of the relevant decisions. Each one of the above enumerated topics constitutes a vast research field on its own, demanding extensive

© The Author(s), under exclusive license to Springer Nature Switzerland AG 2021   83
D. Souravlias et al., *Algorithm Portfolios*, SpringerBriefs in Optimization,
https://doi.org/10.1007/978-3-030-68514-0_7

experience and insight from the practitioner's side. This is counterbalanced by the obtained performance that, under proper setting, can be exceptional, outperforming all the constituent algorithms of the portfolio.

Within this framework, we reviewed the backbone of algorithm portfolio design. Focusing on the dominant metaheuristics that we reviewed in the first chapter, we introduced the reader to the basics of algorithm portfolios and some recent developments in the second chapter. The third chapter outlined methods for selecting the constituent algorithms, including feature-based and statistical selection procedures. Sophisticated resource allocation mechanisms were reported in the fourth chapter, mostly reflecting the authors' previous research results. Parallelization issues for efficient implementations in difficult problems were discussed in the fifth chapter, along with a review of relevant models.

Eventually, three recent applications of algorithm portfolios in the fields of combinatorics, production planning, and humanitarian logistics were demonstrated in the sixth chapter. The presented descriptions aimed to shed light on details and design decisions required to successfully apply an algorithm portfolio.

Besides the aforementioned open topics in algorithm portfolios, their integration with other artificial intelligence and machine learning methods, as well as with the whole algorithmic artillery of classical optimization, are expected to offer fascinating new research results. It is our hope that the present book can serve as the Ariadne's thread for researchers and practitioners aiming at a compendious exposure of the field.

References

1. Abounacer, R., Rekik, M., Renaud, J.: An exact solution approach for multi-objective location–transportation problem for disaster response. Comput. Oper. Res. **41**, 83–93 (2014)
2. Adam, S.P., Alexandropoulos, S.A.N., Pardalos, P.M., Vrahatis, M.N.: No free lunch theorem: a review. In: Demetriou, I., Pardalos, P.M. (eds.) Approximation and Optimization, pp. 57–82. Springer, Cham (2019)
3. Adin, R., Epstein, L., Strassler, Y.: The Classification of Circulant Weighing Matrices of Weight 16 and Odd Order. arXiv preprint math/9910164 (1999)
4. Ahmadi, M., Seifi, A., Tootooni, B.: A humanitarian logistics model for disaster relief operation considering network failure and standard relief time: a case study on San Francisco district. Transp. Res. Part E: Logist. Transp. Rev. **75**, 145–163 (2015)
5. Akay, R., Basturk, A., Kalinli, A., Yao, X.: Parallel population-based algorithm portfolios: an empirical study. Neurocomputing **247**, 115–125 (2017)
6. Alba, E.: Parallel Metaheuristics: A New Class of Algorithms. Wiley-Interscience (2005)
7. Almakhlafi, A., Knowles, J.: Systematic construction of algorithm portfolios for a maintenance scheduling problem. In: IEEE Congress on Evolutionary Computation, Cancun, pp. 245–252 (2013)
8. Ananth, G.Y., Kumar, V., Pardalos, P.M.: Parallel processing of discrete optimization problems. Encycl. Microcomput. **13**, 129–147 (1993)
9. Ang, M., Arasu, K., Ma, S., Strassler, Y.: Study of proper circulant weighing matrices with weigh 9. Discret. Math. **308**, 2802–2809 (2008)
10. Aragón Artacho, F.J., Campoy, R., Kotsireas, I., Tam, M.K.: A feasibility approach for constructing combinatorial designs of circulant type. J. Comb. Optim. **35**(4), 1061–1085 (2018)
11. Arasu, K., Dillon, J., Jungnickel, D., Pott, A.: The solution of the Waterloo problem. J. Comb. Theory Ser. A **71**, 316–331 (1995)
12. Arasu, K., Gulliver, T.: Self-dual codes over F_p and weighing matrices. IEEE Trans. Inf. Theory **47**(5), 2051–2055 (2001)
13. Arasu, K., Gutman, A.: Circulant weighing matrices. Cryptogr. Commun. **2**, 155–171 (2010)
14. Arasu, K., Leung, K., Ma, S., Nabavi, A., Ray-Chaudhuri, D.: Determination of all possible orders of weight 16 circulant weighing matrices. Finite Fields Appl. **12**, 498–538 (2006)
15. Arasu, K.T., Bayes, K., Nabavi, A.: Nonexistence of two circulant weighing matrices of weight 81. Trans. Comb. **4**(3), 43–52 (2015)

16. Arasu, K.T., Dillon, J.F.: Perfect ternary arrays. In: Difference Sets, Sequences and Their Correlation Properties (Bad Windsheim, 1998). NATO Advanced Science Institutes Series, Series C, Mathematical and Physical Sciences, vol. 542, pp. 1–15. Kluwer Academic Publishers (1999)

17. Arasu, K.T., Gordon, D.M., Zhang, Y.: New Nonexistence Results on Circulant Weighing Matrices. arXiv preprint arXiv:1908.08447 (2019)

18. Arasu, K.T., Kotsireas, I.S., Koukouvinos, C., Seberry, J.: On circulant and two-circulant weighing matrices. Australas. J. Comb. **48**, 43–51 (2010)

19. Arasu, K.T., Leung, K.H., Ma, S.L., Nabavi, A., Ray-Chaudhuri, D.K.: Circulant weighing matrices of weight 2^{2t}. Designs Codes Cryptogr. Int. J. **41**(1), 111–123 (2006)

20. Arasu, K.T., Leung, K.H., Ma, S.L., Nabavi, A., Ray-Chaudhuri, D.K.: Determination of all possible orders of weight 16 circulant weighing matrices. Finite Fields Appl. **12**(4), 498–538 (2006)

21. Arasu, K.T., Ma, S.L.: Some new results on circulant weighing matrices. J. Algebraic Comb. **14**(2), 91–101 (2001)

22. Arasu, K.T., Ma, S.L.: Nonexistence of CW(110, 100). Designs Codes Cryptogr. **62**(3), 273–278 (2012)

23. Arasu, K.T., Nabavi, A.: Nonexistence of CW(154, 36) and CW(170, 64). Discret. Math. **311**(8–9), 769–779 (2011)

24. Arasu, K.T., Seberry, J.: On circulant weighing matrices. Australas. J. Comb. **17**, 21–37 (1998)

25. Arasu, K.T., Torban, D.: New weighing matrices of weight 25. J. Comb. Designs **7**(1), 11–15 (1999)

26. Auger, A., Hansen, N.: A restart CMA evolution strategy with increasing population size. In: IEEE Congress on Evolutionary Computation, Edinburgh, pp. 1769–1776 (2005)

27. Banomyong, R., Varadejsatitwong, P., Oloruntoba, R.: A systematic review of humanitarian operations, humanitarian logistics and humanitarian supply chain performance literature 2005 to 2016. Ann. Oper. Res. **283**(1–2), 71–86 (2019)

28. Battiti, R., Mascia, F.: An algorithm portfolio for the sub-graph isomorphism problem. In: Stützle, T., Birattari, M., Hoos, H.H. (eds.) Engineering Stochastic Local Search Algorithms. Designing, Implementing and Analyzing Effective Heuristics, International Workshop, SLS. Lecture Notes in Computer Science, vol. 4638, pp. 106–120. Springer, Berlin, Heidelberg (2007)

29. Birattari, M., Yuan, Z., Balaprakash, P., Stützle, T.: F-race and iterated F-race: an overview. In: Bartz-Beielstein, T., Chiarandini, M., Paquete, L., Preuss, M. (eds.) Experimental Methods for the Analysis of Optimization Algorithms, pp. 311–336. Springer, Berlin, Heidelberg (2010)

30. Borwein, J.M., Sims, B.: The Douglas-Rachford algorithm in the absence of convexity. In: Fixed-Point Algorithms for Inverse Problems in Science and Engineering. Springer Optimization and Its Applications, vol. 49, pp. 93–109. Springer, New York (2011)

31. Brahimi, N., Absi, N., Dauzère-Pérès, S., Nordli, A.: Single-item dynamic lot-sizing problems: an updated survey. Eur. J. Oper. Res. **263**(3), 838–863 (2017)

32. Brahimi, N., Dauzère-Pérès, S., Najid, N.M., Nordli, A.: Single item lot sizing problems. Eur. J. Oper. Res. **168**(1), 1–16 (2006)

33. Brandão, J.: Iterated local search algorithm with ejection chains for the open vehicle routing problem with time windows. Comput. Ind. Eng. **120**, 146–159 (2018)

34. Calderín, J.F., Masegosa, A.D., Pelta, D.A.: An algorithm portfolio for the dynamic maximal covering location problem. Memetic Comput. **9**, 141–151 (2016)

35. Chiarandini, M., Kotsireas, I., Koukouvinos, C., Paquete, L.: Heuristic algorithms for Hadamard matrices with two circulant cores. Theor. Comput. Sci. **407**(1–3), 274–277 (2008)

36. Clerc, M.: Particle Swarm Optimization, vol. 93. Wiley (2010)

37. Clerc, M., Kennedy, J.: The particle swarm–explosion, stability, and convergence in a multidimensional complex space. IEEE Trans. Evol. Comput. **6**(1), 58–73 (2002)

38. Cousineau, J., Kotsireas, I., Koukouvinos, C.: Genetic algorithms for orthogonal designs. Australas. J. Comb. **35**, 263–272 (2006)

39. Circulant Weighing Matrices: https://www.dmgordon.org/cwm/. Accessed 18 July 2020
40. van Dam, W.: Quantum algorithms for weighing matrices and quadratic residues. Algorithmica **34**, 413–428 (2002)
41. Das, S., Mullick, S., Suganthan, P.: Recent advances in differential evolution – an updated survey. Swarm Evol. Comput. **27**, 1–30 (2016)
42. Das, S., Suganthan, P.: Differential evolution: a survey of the state-of-the-art. IEEE Trans. Evol. Comput. **15**(1), 4–31 (2011)
43. Defryn, C., Kenneth, S.: A fast two-level variable neighborhood search for the clustered vehicle routing problem. Comput. Oper. Res. **83**, 78–94 (2017)
44. Eades, P.: On the existence of orthogonal designs. Ph.D. thesis, Australian National University, Canberra (1997)
45. Eades, P., Hain, R.: On circulant weighing matrices. Ars Comb. **2**, 265–284 (1976)
46. Ferreira, A., Pardalos, P.M.: Solving combinatorial optimization problems in parallel-methods and techniques. In: Lecture Notes in Computer Science, vol. 1054. Springer, Berlin, Heidelberg (1996)
47. Ferrer, J.M., Ortuño, M.T., Tirado, G.: A GRASP metaheuristic for humanitarian aid distribution. J. Heuristics **22**(1), 55–87 (2016)
48. Gagliolo, M., Schmidhuber, J.: Algorithm portfolio selection as a bandit problem with unbounded losses. Ann. Math. Artif. Intell. **61**(2), 49–86 (2011)
49. Gagliolo, M., Zhumatiy, V., Schmidhuber, J.: Adaptive online time allocation to search algorithms. In: European Conference on Machine Learning, Pisa, pp. 134–143 (2004)
50. García, S., Fernández, A., Luengo, J., Herrera, F.: Advanced nonparametric tests for multiple comparisons in the design of experiments in computational intelligence and data mining: experimental analysis of power. Inf. Sci. **180**(10), 2044–2064 (2010)
51. Gendreau, M., Potvin, J.Y.: Tabu search. In: Gendreau, M., Potvin, J.Y. (eds.) Handbook of Metaheuristics, pp. 41–59. Springer, Boston (2010)
52. Gendreau, M., Potvin, J.Y.: Handbook of Metaheuristics, vol. 272. Springer (2019)
53. Geramita, A.V., Seberry, J.: Orthogonal designs: quadratic forms and hadamard matrices. Marcel Dekker, New York, Basel, p. viii, 460 (1979)
54. Glover, F.: Tabu search – part I. ORSA J. Comput. **1**, 190–206 (1989)
55. Glover, F.: Tabu search – part II. ORSA J. Comput. **2**, 4–32 (1990)
56. Glover, F., Laguna, M.: Tabu Search. Kluwer Academic Publishers, Norwell (1997)
57. Golany, B., Yang, J., Yu, G.: Economic lot-sizing with remanufacturing options. IIE Trans. **33**(11), 995–1004 (2001)
58. Gomes, C.P., Selman, B.: Algorithm portfolios. Artif. Intell. **126**(1–2), 43–62 (2001)
59. Gong, W., Cai, Z., Ling, C.X., Li, H.: Enhanced differential evolution with adaptive strategies for numerical optimization. IEEE Trans. Syst. Man Cybern. Part B (Cybern.) **41**(2), 397–413 (2010)
60. Gralla, E., Goentzel, J.: Humanitarian transportation planning: evaluation of practice-based heuristics and recommendations for improvement. Eur. J. Oper. Res. **269**(2), 436–450 (2018)
61. Gropp, W., Gropp, W.D., Lusk, E., Lusk, A.D.F.E.E., Skjellum, A.: Using MPI: portable parallel programming with the message-passing interface, vol. 1. MIT Press (1999)
62. Hain, R.M.: Circulant weighing matrices. Master's thesis, The Australian National University (1977)
63. Hansen, P., Brimberg, J., Urošević, D., Mladenović, N.: Primal-dual variable neighborhood search for the simple plant-location problem. INFORMS J. Comput. **19**(4), 552–564 (2007)
64. Hansen, P., Mladenović, N., Brimberg, J., Pérez, J.A.M.: Variable neighborhood search. In: Gendreau, M., Potvin, J. (eds.) Handbook of Metaheuristics, pp. 57–97. Springer (2019)
65. Hansen, P., Mladenović, N., Todosijević, R., Hanafi, S.: Variable neighborhood search: basics and variants. EURO J. Comput. Optim. **5**(3), 423–454 (2017)
66. He, Y., Yuen, S.Y., Lou, Y., Zhang, X.: A sequential algorithm portfolio approach for black box optimization. Swarm Evol. Comput. **44**, 559–570 (2019)
67. Holguín-Veras, J., Jaller, M., Wassenhove, L.N.V., Pérez, N., Wachtendorf, T.: On the unique features of post-disaster humanitarian logistics. J. Oper. Manag. **30**(7), 494–506 (2012)

68. Huberman, B.A., Lukose, R.M., Hogg, T.: An economics approach to hard computational problems. Science **27**, 51–53 (1997)
69. Hyndman, R.J.: Moving Averages. Springer, Berlin, Heidelberg (2011)
70. Jarboui, B., Siarry, P., Teghem, J., Bourrieres, J.P.: Metaheuristics for Production Scheduling. Wiley Online Library (2013)
71. Jordon, J., Yoon, J., van der Schaar, M.: Knockoffgan: generating knockoffs for feature selection using generative adversarial networks. In: International Conference on Learning Representations, Vancouver (2018)
72. Karaboga, D., Basturk, B.: A powerful and efficient algorithm for numerical function optimization: artificial bee colony (ABC) algorithm. J. Glob. Optim. **39**(3), 459–471 (2007)
73. Kennedy, J.: Small worlds and mega-minds: effects of neighborhood topology on particle swarm performance. In: IEEE Congress on Evolutionary Computation, Washington, DC, vol. 3, pp. 1931–1938 (1999)
74. Kennedy, J., Eberhart, R.C.: Particle swarm optimization. In: IEEE International Conference on Neural Networks, Piscataway, vol. IV, pp. 1942–1948 (1995)
75. Kerschke, P., Hoos, H., Neumann, F., Trautmann, H.: Automated algorithm selection: survey and perspectives. Evol. Comput. **27**(1), 3–45 (2019)
76. Korkou, T., Souravlias, D., Parsopoulos, K., Skouri, K.: Metaheuristic optimization for logistics in natural disasters. In: International Conference on Dynamics of Disasters, Kalamata, pp. 113–134 (2016)
77. Kotsireas, I.: Algorithms and metaheuristics for combinatorial matrices. In: Pardalos, P., Du, D.Z., Graham, R.L. (eds.) Handbook of Combinatorial Optimization, pp. 283–309. Springer, New York (2013)
78. Kotsireas, I., Koukouvinos, C., Pardalos, P., Simos, D.: Competent genetic algorithms for weighing matrices. J. Comb. Optim. **24**(4), 508–525 (2012)
79. Kotsireas, I., Parsopoulos, K., Piperagkas, G., Vrahatis, M.: Ant-based approaches for solving autocorrelation problems. In: Dorigo, M., et al. (eds.) Lecture Notes in Computer Science, vol. 7461, pp. 220–227. Springer, Berlin, Heidelberg (2012)
80. Kotsireas, I.S., Pardalos, P.M., Parsopoulos, K.E., Souravlias, D.: On the solution of circulant weighing matrices problems using algorithm portfolios on multi-core processors. In: International Symposium on Experimental Algorithms, St. Petersburg, pp. 184–200 (2016)
81. Kotthoff, L.: Algorithm selection for combinatorial search problems: a survey. In: Bessiere, C., et al. (eds.) Data Mining and Constraint Programming: Foundations of a Cross-Disciplinary Approach, pp. 149–190. Springer, Cham (2016)
82. Koukouvinos, C., Seberry, J.: Weighing matrices and their applications. J. Statist. Plann. Inference **62**(1), 91–101 (1997)
83. Kovács, G., Spens, K.M.: Relief Supply Chain Management for Disasters: Humanitarian, Aid and Emergency Logistics. IGI Global (2012)
84. Leung, K.H., Ma, S.L.: Proper circulant weighing matrices of weight 25. preprint (2011)
85. Leung, K.H., Schmidt, B.: Finiteness of circulant weighing matrices of fixed weight. J. Comb. Theory Ser. A **118**(3), 908–919 (2011)
86. Leung, K.H., Schmidt, B.: Structure of group invariant weighing matrices of small weight. J. Comb. Theory Ser. A **154**, 114–128 (2018)
87. Leyton-Brown, K., Nudelman, E., Andrew, G., McFadden, J., Shoham, Y.: A portfolio approach to algorithm selection. In: International Joint Conference on Artificial Intelligence, Acapulco, pp. 1542–1543 (2003)
88. Lindauer, M., Hoos, H., Hutter, F.: From sequential algorithm selection to parallel portfolio selection. In: International Conference on Learning and Intelligent Optimization, Lille, pp. 1–16 (2015)
89. Lindauer, M., Hoos, H., Leyton-Brown, K., Schaub, T.: Automatic construction of parallel portfolios via algorithm configuration. Artif. Intell. **244**, 272–290 (2017)
90. Liu, S., Tang, K., Yao, X.: Generative adversarial construction of parallel portfolios. IEEE Trans. Cybern. 1–12 (2020)

91. Loreggia, A., Malitsky, Y., Samulowitz, H., Saraswat, V.: Deep learning for algorithm portfolios. In: Thirtieth AAAI Conference on Artificial Intelligence, Phoenix, pp. 1280–1286 (2016)

92. Lourenço, H.R., Martin, O.C., Stützle, T.: Iterated local search. In: Glover, F., Kochenberger, G.A. (eds.) Handbook of Metaheuristics, pp. 320–353. Springer, US (2003)

93. Lourenço, H.R., Martin, O.C., Stützle, T.: Iterated local search: framework and applications. In: Gendreau, M., Potvin, J. (eds.) Handbook of Metaheuristics, pp. 129–168. Springer (2019)

94. Lucas, J.M., Saccucci, M.S.: Exponentially weighted moving average control schemes: properties and enhancements. Technometrics $32(1)$, 1–12 (1990)

95. Ma, W., Wang, M., Zhu, X.: Improved particle swarm optimization based approach for bilevel programming problem-an application on supply chain model. Int. J. Mach. Learn. Cybern. $5(2)$, 281–292 (2014)

96. Makridakis, S., Wheelwright, S.C., Hyndman, R.J.: Forecasting Methods and Applications. Wiley (2008)

97. Menéndez, B., Pardo, E.G., Alonso-Ayuso, A., Molina, E., Duarte, A.: Variable neighborhood search strategies for the order batching problem. Comput. Oper. Res. 78, 500–512 (2017)

98. Meng, Z., Pan, J.S., Tseng, K.K.: PaDE: an enhanced differential evolution algorithm with novel control parameter adaptation schemes for numerical optimization. Knowl.-Based Syst. 168, 80–99 (2019)

99. Mladenovic, N., Hansen, P.: Variable neighborhood search. Comput. Oper. Res. $24(11)$, 1097–1100 (1997)

100. Mohamed, A.W.: RDEL: restart differential evolution algorithm with local search mutation for global numerical optimization. Egypt. Inform. J. $15(3)$, 175–188 (2014)

101. Mohamed, A.W.: An improved differential evolution algorithm with triangular mutation for global numerical optimization. Comput. Ind. Eng. 85, 359–375 (2015)

102. Moreno, A., Alem, D., Ferreira, D.: Heuristic approaches for the multiperiod location-transportation problem with reuse of vehicles in emergency logistics. Comput. Oper. Res. 69, 79–96 (2016)

103. Moustaki, E., Parsopoulos, K.E., Konstantaras, I., Skouri, K., Ganas, I.: A first study of particle swarm optimization on the dynamic lot sizing problem with product returns. In: XI Balkan Conference on Operational Research (BALCOR 2013), Belgrade, pp. 348–356 (2013)

104. Muñoz, M., Kirley, M.: ICARUS: identification of complementary algorithms by uncovered sets. In: IEEE Congress on Evolutionary Computation, Vancouver, pp. 2427–2432 (2016)

105. Muñoz, M., Sun, Y., Kirley, M., Halgamuge, S.: Algorithm selection for black-box continuous optimization problems: a survey on methods and challenges. Inf. Sci. 317, 224–245 (2015)

106. Nabavi, A.: The spectrum of circulant weighing matrices of weight 16. ProQuest LLC, Ann Arbor (2000). Thesis (Ph.D.)–The Ohio State University

107. Nagurney, A., Qiang, Q.: Quantifying supply chain network synergy for humanitarian organizations. IBM J. Res. Dev. $64(1/2)$, 12:1–12:16 (2020)

108. Nasr, N., Thurston, M.: Remanufacturing: A Key Enabler to Sustainable Product Systems, pp. 15–18. Rochester Institute of Technology (2006)

109. Nesmachnow, S.: An overview of metaheuristics: accurate and efficient methods for optimisation. Int. J. Metaheuristics $3(4)$, 320–347 (2014)

110. Onar, S.Ç., Öztayşi, B., Kahraman, C., Yanık, S., Şenvar, Ö.: A literature survey on metaheuristics in production systems. In: Talbi, E., Yalaoui, F., Amodeo, L. (eds.) Metaheuristics for Production Systems, pp. 1–24. Springer, Cham (2016)

111. Özdamar, L., Ekinci, E., Kucukyazici, B.: Emergency logistics planning in natural disasters. Ann. Oper. Res. $129(1–4)$, 217–245 (2004)

112. Özdamar, L., Ertem, M.A.: Models, solutions and enabling technologies in humanitarian logistics. Eur. J. Oper. Res. $244(1)$, 55–65 (2015)

113. Pardalos, P.M.: Parallel processing of discrete problems. In: IMA Volumes in Mathematics and Its Applications, vol. 106. Springer, New York (1999)

114. Pardalos, P.M., Phillips, A.T., Rosen, J.B.: Topics in Parallel Computing in Mathematical Programming, vol. 2. American Mathematical Society & Science Press (1992)

115. Pardalos, P.M., Rajasekaran, S.: Advances in randomized parallel computing. Kluwer Academic, US (1999)
116. Parsopoulos, K.E., Konstantaras, I., Skouri, K.: Metaheuristic optimization for the single-item dynamic lot sizing problem with returns and remanufacturing. Comput. Ind. Eng. **83**, 307–315 (2015)
117. Parsopoulos, K.E., Vrahatis, M.N.: Particle Swarm Optimization and Intelligence: Advances and Applications. Information Science Publishing (IGI Global) (2010)
118. Peng, F., Tang, K., Chen, G., Yao, X.: Population-based algorithm portfolios for numerical optimization. IEEE Trans. Evol. Comput. **14**(5), 782–800 (2010)
119. Petalas, Y.G., Parsopoulos, K.E., Vrahatis, M.N.: Improving fuzzy cognitive maps learning through memetic particle swarm optimization. Soft Comput. **13**(1), 77–94 (2009)
120. Pigosso, D.C., Zanette, E.T., Guelere Filho, A., Ometto, A.R., Rozenfeld, H.: Ecodesign methods focused on remanufacturing. J. Clean. Prod. **18**(1), 21–31 (2010)
121. Piñeyro, P., Viera, O.: The economic lot-sizing problem with remanufacturing and one-way substitution. Int. J. Prod. Econ. **124**(2), 482–488 (2010)
122. Piñeyro, P., Viera, O.: The economic lot-sizing problem with remanufacturing: analysis and an improved algorithm. J. Remanuf. **5**, 1–13 (2015)
123. Piperagkas, G.S., Konstantaras, I., Skouri, K., Parsopoulos, K.E.: Solving the stochastic dynamic lot-sizing problem through nature-inspired heuristics. Comput. Oper. Res. **39**(7), 1555–1565 (2012)
124. Piperagkas, G.S., Voglis, C., Tatsis, V.A., Parsopoulos, K.E., Skouri, K.: Applying PSO and DE on multi-item inventory problem with supplier selection. In: 9th Metaheuristics International Conference, Udine, pp. 359–368 (2011)
125. Poli, R.: An Analysis of Publications on Particle Swarm Optimisation Applications. Technical Report CSM-649, University of Essex, Department of Computer Science (2007)
126. Rana, S., Jasola, S., Kumar, R.: A boundary restricted adaptive particle swarm optimization for data clustering. Int. J. Mach. Learn. Cybern. **4**(4), 391–400 (2013)
127. Rauchecker, G., Schryen, G.: An exact branch-and-price algorithm for scheduling rescue units during disaster response. Eur. J. Oper. Res. **272**(1), 352–363 (2019)
128. Resende, M.G., Martí, R., Pardalos, P.: Handbook of Heuristics. Springer (2017)
129. Retel Helmrich, M.J., Jans, R., van den Heuvel, W., Wagelmans, A.P.: Economic lot-sizing with remanufacturing: complexity and efficient formulations. IIE Trans. **46**(1), 67–86 (2014)
130. Ribeiro, C., Resende, M.: Path-relinking intensification methods for stochastic local search algorithms. J. Heuristics **18**(2), 193–214 (2012)
131. Rice, J.R.: The algorithm selection problem. Adv. Comput. **15**, 65–118 (1976)
132. Richey, R.G., Kovács, G., Spens, K.: Identifying challenges in humanitarian logistics. Int. J. Physical Distrib. Logist. Manag. **39**(6), 506–528 (2009)
133. Rivera, J.C., Afsar, H.M., Prins, C.: Mathematical formulations and exact algorithm for the multitrip cumulative capacitated single-vehicle routing problem. Eur. J. Oper. Res. **249**(1), 93–104 (2016)
134. Roberts, S.W.: Control chart tests based on geometric moving averages. Technometrics **1**(3), 239–250 (1959)
135. Schmidt, B., Smith, K.W.: Circulant weighing matrices whose order and weight are products of powers of 2 and 3. J. Comb. Theory Ser. A **120**(1), 275–287 (2013)
136. Schulz, T.: A new silver–meal based heuristic for the single–item dynamic lot sizing problem with returns and remanufacturing. Int. J. Prod. Res. **49**(9), 2519–2533 (2011)
137. Shukla, N., Dashora, Y., Tiwari, M., Chan, F., Wong, T.: Introducing algorithm portfolios to a class of vehicle routing and scheduling problem. In: Operations and Supply Chain Management (OSCM 2007), Bangkok, pp. 1015–1026 (2007)
138. Sifaleras, A., Konstantaras, I., Mladenović, N.: Variable neighborhood search for the economic lot sizing problem with product returns and recovery. Int. J. Prod. Econ. **160**, 133–143 (2015)
139. Song, T., Liu, S., Tang, X., Peng, X., Chen, M.: An iterated local search algorithm for the University Course Timetabling Problem. Appl. Soft Comput. **68**, 597–608 (2018)

140. Souravlias, D., Kotsireas, I.S., Pardalos, P.M., Parsopoulos, K.E.: Parallel algorithm portfolios with performance forecasting. Optim. Methods Softw. **34**(6), 1231–1250 (2019)
141. Souravlias, D., Parsopoulos, K.E.: On the design of metaheuristics-based algorithm portfolios. In: Pardalos, P.M., Migdalas, A. (eds.) Open Problems in Optimization and Data Analysis, pp. 271–284. Springer, Cham (2018)
142. Souravlias, D., Parsopoulos, K.E., Alba, E.: Parallel algorithm portfolio with market trading-based time allocation. In: Lübbecke, M., et al. (eds.) Operations Research Proceedings 2014, pp. 567–574. Springer, Switzerland (2016)
143. Souravlias, D., Parsopoulos, K.E., Kotsireas, I.S.: Circulant weighing matrices: a demanding challenge for parallel optimization metaheuristics. Optim. Lett. **10**(6), 1303–1314 (2016)
144. Souravlias, D., Parsopoulos, K.E., Meletiou, G.C.: Designing bijective S-boxes using algorithm portfolios with limited time budgets. Appl. Soft Comput. **59**, 475–486 (2017)
145. Storn, R., Price, K.: Differential evolution–a simple and efficient heuristic for global optimization over continuous spaces. J. Glob. Optim. **11**(4), 341–359 (1997)
146. Strassler, Y.: The classification of circulant weighing matrices of weight 9. Ph.D. thesis, Bar-Ilan University (1997)
147. Strassler, Y.: New circulant weighing matrices of prime order in CW(31, 16), CW(71, 25), CW(127, 64). J. Statist. Plann. Inference **73**(1–2), 317–330 (1998)
148. Talbi, E.G., Yalaoui, F., Amodeo, L.: Metaheuristics for Production Systems. Springer (2016)
149. Tan, M.M.: Relative difference sets and cw matrices. Ph.D. thesis, Nanyang Technological University, Singapore (2014)
150. Tan, M.M.: Group invariant weighing matrices. Designs Codes Cryptogr. Int. J. **86**(12), 2677–2702 (2018)
151. Tang, K., Peng, F., Chen, G., Yao, X.: Population-based algorithm portfolios with automated constituent algorithms selection. Inf. Sci. **279**, 94–104 (2014)
152. Tasgetiren, F., Chen, A., Gencyilmaz, G., Gattoufi, S.: Smallest position value approach. Stud. Comput. Intell. **175**, 121–138 (2009)
153. Tatham, P., Christopher, M.: Humanitarian logistics: meeting the challenge of preparing for and responding to disasters. Kogan Page Publishers (2018)
154. Teunter, R.H., Bayindir, Z.P., Van den Heuvel, W.: Dynamic lot sizing with product returns and remanufacturing. Int. J. Prod. Res. **44**(20), 4377–4400 (2006)
155. Thomas, A., Kopczak, L.: From Logistics to Supply Chain Management – The Path Forward to the Humanitarian Sector. Fritz Institute (2005)
156. Tian, N., Lai, C.H.: Parallel quantum-behaved particle swarm optimization. Int. J. Mach. Learn. Cybern. **5**(2), 309–318 (2014)
157. Tofighi, S., Torabi, S.A., Mansouri, S.A.: Humanitarian logistics network design under mixed uncertainty. Eur. J. Oper. Res. **250**(1), 239–250 (2016)
158. Tong, H., Liu, J., Yao, X.: Algorithm portfolio for individual-based surrogate-assisted evolutionary algorithms. In: Genetic and Evolutionary Computation Conference, Prague, pp. 943–950 (2019)
159. Van Wassenhove, L.N.: Humanitarian aid logistics: supply chain management in high gear. J. Oper. Res. Soc. **57**(5), 475–489 (2006)
160. Vrugt, J.A., Robinson, B.A., Hyman, J.M.: Self-adaptive multimethod search for global optimization in real-parameter spaces. IEEE Trans. Evol. Comput. **13**(2), 243–259 (2009)
161. Wagner, H.M., Whitin, T.M.: Dynamic version of the economic lot size model. Manag. Sci. **5**(1), 88–96 (1958)
162. Wang, X., Choi, T.M., Liu, H., Yue, X.: A novel hybrid ant colony optimization algorithm for emergency transportation problems during post-disaster scenarios. IEEE Trans. Syst. Man Cybern. Syst. **48**(4), 545–556 (2016)
163. Wang, X.Z., He, Y.L., Dong, L.C., Zhao, H.Y.: Particle swarm optimization for determining fuzzy measures from data. Inf. Sci. **181**(19), 4230–4252 (2011)
164. Wawrzyniak, J., Drozdowski, M., Sanlaville, É.: Selecting algorithms for large berth allocation problems. Eur. J. Oper. Res. **283**(3), 844–862 (2020)

165. Wilcoxon, F., Katti, S., Wilcox, R.A.: Critical values and probability levels for the Wilcoxon rank sum test and the Wilcoxon signed rank test. Sel. Tables Math. Statist. **1**, 171–259 (1970)

166. Wolpert, D.H., Macready, W.G.: No free lunch theorem for optimization. IEEE Trans. Evol. Comput. **1**, 67–82 (1997)

167. Xu, L., Hutter, F., Hoos, H., Leyton-Brown, K.: SATzilla: portfolio-based algorithm selection for SAT. J. Artif. Intell. Res. **32**(1), 565–606 (2008)

168. Yang, S., MR, A.R., Kaminski, J., Pepin, H.: Opportunities for industry 4.0 to support remanufacturing. Appl. Sci. **8**(7), 1177 (2018)

169. Yeniay, Ö.: Penalty function methods for constrained optimization with genetic algorithms. Math. Comput. Appl. **10**(1), 45–56 (2005)

170. Yorgov, V.: On the existance of certain circulant weighing matrices. J. Comb. Math. Comb. Comput. **86**, 73–85 (2013)

171. Yu, W., Shen, M., Chen, W., Zhan, Z., Gong, Y., Lin, Y., Liu, O., Zhang, J.: Differential evolution with two-level parameter adaptation. IEEE Trans. Cybern. **44**(7), 1080–1099 (2014)

172. Yuen, S., Chow, C., Zhang, X., Lou, Y.: Which algorithm should I choose: an evolutionary algorithm portfolio approach. Appl. Soft Comput. **40**, 654–673 (2016)

173. Yuen, S.Y., Zhang, X.: On composing an algorithm portfolio. Memetic Comput. **7**, 203–214 (2015)

174. Yun, X., Epstein, S.L.: Learning algorithm portfolios for parallel execution. In: Hamadi, Y., Schoenauer, M. (eds.) Learning and Intelligent Optimization. LION 2012. Lecture Notes in Computer Science, vol. 7219, pp. 323–338. Springer, Berlin, Heidelberg (2012)

175. Zhu, Q., Li, H., Zhao, S., Lun, V.: Redesign of service modes for remanufactured products and its financial benefits. Int. J. Prod. Econ. **171**, 231–240 (2016)

176. Zobolas, G.I., Tarantilis, C.D., Ioannou, G.: Minimizing makespan in permutation flow shop scheduling problems using a hybrid metaheuristic algorithm. Comput. Operat. Res. **36**, 1249–1267 (2009)

177. Zohali, H., Naderi, B., Mohammadi, M., Roshanaei, V.: Reformulation, linearization, and a hybrid iterated local search algorithm for economic lot-sizing and sequencing in hybrid flow shop problems. Comput. Oper. Res. **104**, 127–138 (2019)